Calculus For Cats

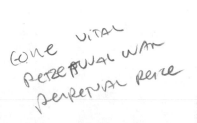

Calculus for Cats

by
Kenn Amdahl and Jim Loats, Ph.D

Calculus for Cats
copyright 2001 by Kenn Amdahl and Jim Loats, Ph.D

Cover drawing by Carol McKellor, who reserves all rights.
Cover photograph courtesy of Alex Schumacher

Published by Clearwater Publishing Company, Inc.
PO Box 778
Broomfield, CO 80038

(303) 436-1982

Acknowledgements

This was not an easy project. As you may remember from *Algebra Unplugged*, Kenn does not particularly like or trust math, and when confronted with the task of sitting down and learning the concepts of calculus, he was easily and gratefully distracted. Kenn loves words, literature, and language. But not math.

Jim, on the other hand, adores math. To him, there are few things as fun as filling up blank pages with mathematical scribblings. While he was in mathematical nirvana, poking at his calculator or burning up a number two pencil, he didn't always notice Kenn sneaking out the back door to read a book. Jim has not spent his life reading or writing. He's not someone who salivates over words. So, when Kenn started talking about the writing process, Jim's mind tip-toed off to his sailboat.

To write this book, the guy who loves words (but distrusts math) and the guy who loves math (but distrusts words) were metaphorically chained together for over three years. The gist of many conversations was an astonished, "You've never read Steinbeck?" followed by the equally astonished response, "You don't know the Pythagorean Theorem?" Now, these are both big guys, each over six feet tall and over two hundred pounds. Neither one is shy; both are articulate, reasonably bright and unreasonably stubborn, with good, strong voices. More than one coffee shop waitress edged backward, away from their calculus discussions, with the nervous delicacy of a bartender easing toward his baseball bat.

Reality television isn't ready for this show.

You need to know all that to appreciate how much we'd like to thank all the folks who lived with us while we were writing the book. Cheryl Amdahl and Dee Loats deserve special medals. Our friends, offspring, musical partners, and co-workers also deserve our thanks.

Specific thanks to the following: Carol McKellor who drew the kitty for the cover; Alex Schumacher for the use of the tiger picture on the cover; Alex Goulder for editing; Cynthia Holcomb for her comments and suggestions; Keith Devlin for his back cover comment; Scott Amdahl and Jeff Loats for their editorial suggestions; the math department at Metropolitan State College in Denver; and Martin Gardner for suggesting we write the book in the first place.

After the book was finished, Jim invited Kenn and his wife onto his new sailboat. It was a pleasant time among good friends, with no talk of either math or literature. We hope your experience reading *Calculus for Cats* is just like that: an easy sail through fine weather.

Contents

Integral Calculus

Extra Stuff

About this book

Most calculus books resemble an ambitious Tolstoy novel in size and weight, but contain very few actual words: it takes about two hours to read one from cover to cover if you only read the verbal explanations and skip all the exercises, symbols, proofs, and sample problems. Unfortunately, should you decide to read one that way, you probably won't gain much of an understanding about calculus, because all those examples and exercises are their primary method of conveying the subject.

This book is different. Despite its small size, it contains more words to explain the basic ideas, but no exercises at all. Its purpose is not to replace your Required Calculus Textbook, but to augment the explanations of concepts covered in class. You can read this book without any prior understanding of the subject and come away with a sense of what calculus is, why it's powerful, and when and how to use it. You will not become an expert at the subject – that's not our goal. *Calculus for Cats* is, in the purest sense of the word, an introduction.

It provides an easy overview of the subject for people who never took calculus and want to broaden their horizons, as well as for people who are considering taking a calculus class but are frightened and intimidated by the name alone. Math majors may enjoy the book and find it useful, but we hope English Literature majors will as well. Some poetry instructors feel that it is crucial to understand iambic pentameter before reading a sonnet. Our favorite poetry instructors first let their students experience the power of a great poem, whether by Shakespeare, or Donne, or the latest popular singer. Once a student feels that power, they're hooked and willingly master the mechanical aspects.

We don't claim to be Shakespeares. But we do think the beauty and power of calculus speaks for itself when stripped of its more technical underpinnings. We hope that our simple and sometimes silly explanations don't insult those who understand the beauty that lies beneath the surface. With luck, our less-intimidating format may prepare some students to appreciate that beauty as it unfolds in class.

Math instructors may have to fight down a sense of indignation from time to time as we skip the proofs of why things work, or ignore the logical steps that lead up to certain techniques, and they may cringe at our apparently casual use of terms that have not been rigorously defined. It's not that we don't respect rigor, precision, and comprehensiveness; they just don't describe our mission here. Many excellent calculus text books already address those issues, and provide dozens of examples and exercises. You just can't read them without an instructor to translate them for you.

The goal of this book is to provide additional explanations of concepts. If you think of calculus as a foreign language, this book won't prepare you for a career as a professional translator. It's more a tourist guide book of useful phrases.

Introduction

Approximately four thousand years ago, aliens invaded Earth and began implementing a diabolical plan to enslave humanity, to force us to build their homes, provide them the most expensive and exotic foods, tend their every whim and most trivial wish, regardless of the inconvenience, while they lounged in splendor and did nothing.

These aliens have come to be known as "cats."

The conquest proved simple. Although the creatures lacked opposable thumbs, were inferior in size and had limited capabilities for speech, they had one overwhelmingly superior ability.

They understood calculus.

And humans did not.

Since then, the beasts have employed their considerable cunning to prevent their unwitting subjects from discovering this secret. A few humans have caught on, of course, and have actually learned to use calculus. The aliens have systematically cut these odd ones from the herd and forced them to spend their lives in isolated pens (which are often located in university math departments or cubicles deep within huge corporations). To prevent them from becoming rebellious, they are kept content with huge salaries, extravagant homes and attractive spouses. But the cats watch them carefully. Occasionally, one of these humans will feel the urge to share the secrets, perhaps even write a book.

When this happens, the cats hypnotize the would-be rebel into writing a vague and confusing book filled with incomprehensible diagrams, bizarre symbols, and language adapted from an obscure

Mayan dialect. When confronted with such a volume, most humans simply close the book, choose a new career path, and warm some milk for their sweet little tabby. The aliens win again.

The plan has been wildly successful and the proof is obvious: cats rule the world and very few humans understand calculus.

But now their evil plot has begun to unravel. The brave authors of this book have determined to resist the conquering aliens. We have surrounded ourselves with Dobermans and are prepared to expose this mystery to the world.

Wait! We know what you're thinking.

But before you decide that calculus is beyond you, consider this: if cats can learn it, so can you.

The Catmobile

A catmobile resembles a limousine which has been equipped to the exacting taste of its owners. The seats are crafted of soft pillows and covered with dark, expensive sweaters that must be dry cleaned. Brand new curtains cover the windows in case the imperious passengers feel the need to sharpen their claws. Cats drive the vehicle by casually flicking their tails. You may have seen them practicing this skill.

Cats hide their vehicles well, in crawl-spaces beneath homes, in storm sewers, in thick bushes – anywhere humans are unlikely to look. The entire plot would be at risk if humans knew that cats actually could drive themselves down to the store when they hunger for a snack. Worse, if a human ever understood a catmobile he would also understand calculus. This cannot happen. Those rare times when a human stumbles across one of the little cars, the cats have no choice. They kill the human. Luckily, it seldom happens. *TIME?*

a? A catmobile comes with various interesting gauges: a clock, an odometer, a tilt meter and a status meter. *all measures*

The clock tells cats when it is most inconvenient to request services from their human slaves. This is a constant source of delight for them.

The odometer measures how far they have traveled since their last meal.

The tilt meter shows whether they are traveling up hill, or down hill, or on a level road. The meter might be as simple as a pencil floating in a glass of water, or it might be an elaborate electrical device. There is no way to predict what an individual cat will prefer.

13

The tilt meter reads the exact steepness of whatever slope the catmobile is on at any given instant.

Finally, there is the status meter. Cats don't like to be bothered with checking so many gauges, so the status meter reads each one periodically and tells the cats the status of all of them. It might say, for example, that it's 3:05 PM, we've traveled six miles, and we're going up a 10% incline.

This information tells the riders that it's time to stop and yowl for food.

The status meter can be set to take readings as frequently as the whim of the cats require. It can read the gauges once an hour, or once a minute, or once every tenth of a second.

If you only read your odometer and clock once an hour, you don't know precisely how fast you were moving at any specific instant between measurements. On the other hand, if you've been traveling for exactly one hour and you've gone exactly sixty miles, you can certainly say that you've been moving at an average speed of sixty miles an hour.

That's close enough for most cats. Of course, you might have gone sixty-five miles per hour for a half hour and fifty-five for the next half hour. Cats have a remarkable capacity to ignore anything they don't really care about.

If your tilt meter says you're going up a 15% slope and one minute later it says you're going up a 20% slope, it's easy to picture the indicator needle rising smoothly during that time. Again, that may not be exactly what happened. The road may not rise quite that predictably. To have a more accurate picture, you'd have to monitor the gauges more often.

Unless a can of tuna or a warm lap depends on such precision, a cat will yawn and stretch and say "close enough."

If the tilt meter has been indicating that you're going up a 10% slope for the last several readings, then indicates a 0% slope, then the next several readings say you're going down a 10% slope, the cats might deduce that they crested the hill and are now traveling down the other side. They don't even have to rouse themselves to look out the window. In fact, they can also estimate the exact time they hit the top of the hill. The ability to use the tilt meter is what separates cats from the reptiles, politicians and other lesser creatures. The tilt meter is the key to their entire plot. It's also the topic of this book.

When the cats take a spin in the catmobile and graph the elevation of the road in their evil little brains, they call it "driving." To confuse humans, it has been given the more intimidating name of "differential calculus." Differential calculus is concerned with the changes in the elevation of the road itself, with its slope (or steepness), the distance traveled, and with guessing what happened between measurements. The idea is simple: figure out what the tilt meter reads at any point, then use that to answer questions about the other meters. We assume the road is perfectly straight, it doesn't curve left or right. All we're concerned with is its elevation, the way it goes up and down. Even though the road is straight, if we graph its changing elevation the picture we draw will not necessarily be a straight line. It may curve upward or downward. That is the line we're interested in when we study differential calculus.

Now imagine that the road is on the edge of a perfectly smooth cliff that plummets straight down to the ocean. A mouse dropped from the catmobile window will descend until it splashes into the water. This is great fun.

Even more fun is carefully dipping the mouse in paint before releasing it. Now it has become a squeaking, furry paint brush which makes a straight vertical line on the white cliff before it splashes into the water.

(In deference to mouse lovers everywhere, we point out that these are hypothetical mice who love both painting and swimming and who delight in sliding down cliffs. No mice were actually harmed in creating this analogy. And for those picky physics majors out there, yes, we know that a mouse released from a moving vehicle would have some inertia and so would not fall in a perfectly straight line, but then you have to know some calculus to figure that curve out, and we just got started here, and could you please just give us a break? Our mice fall straight down.)

If we drop mice one right after the other, all these straight paint lines touch each other and the whole cliff gets painted.

Because cats don't like to waste mice in this way, they came up with another amusement to replace it. The game is "how much paint would it take to cover the cliff from where we started to right here if we did drop mice?" Or, as translated by mathematicians: "integral calculus." Integral calculus is concerned not so much with the road itself (the curved line on the graph paper) as it is with the changing area beneath the line.

Amazingly, many things change according to a pattern that is similar to the movement of a catmobile and the mouse art beneath it: profits compared to expenses, the flight paths of cannonballs and missiles, and the strength of bridges as you add metal. Because cats noticed this, they were able to conquer the world.

What is Calculus and Why Is It Terrifying?

Calculus is a tool for understanding things that change in relationship to each other. We use it to express that changing relationship in mathematical terms so we can predict the status of one variable when we know the other, and we use it to determine when the relationship itself changes, and to identify when one of the variables rises to its maximum or falls to its minimum.

We do this every day in less mysterious arenas. As we jog around a track, we notice how long it takes us to run one lap. Almost instinctively we convert this to our "speed." Speed is not a measurement of either time or distance, but the relationship between the two. If we're jogging eight minutes per mile, and we know how far we have to run, we can easily calculate how long it will take us, and what time we'll be done, and whether we'll beat our best time or not. It is much faster to create a simple math problem that uses our speed as one of the variables than it would be to measure each section of track, and note the elapsed time as we pass it. We can probably even do it in our heads. The simple and efficient tactic is to describe the relationship between time and distance and then use that relationship. It's hard to imagine any other way to do it. Without using the concept of "speed" we'd have a hard time making predictions about when we'd finish. We even notice the relationship between time and this new function called speed and call it acceleration. It is not at all uncommon for us to identify the relationship between related variables and use that relationship to deduce new information. Whenever we wonder when we'll get to the next gas station, or how long will it take us to accelerate up to merging speed, we are employing a calculus-like thought process.

We use calculus to express the relationship between more complicated variables. It is especially useful when the two variables are not changing at steady rates. Just as the concept "speed" is a huge short cut which replaces taking hundreds of measurements along a race track, calculus is a huge short cut to answering much more difficult questions. Just as identifying your speed does not answer the question of "when will we finish the race," calculus does not answer questions either. Calculus is only part of the process. We use calculus to describe the relationship, then we use the result in an algebra problem to actually extract the answer to our question. That's what calculus is: a way to express the relationship between changing variables so we can use it in an algebra problem.

Measuring reality with numbers is a tricky business. Most of the world won't sit still for very long, and measuring takes time. By the time we're done measuring a bullet's location, for example, it's already somewhere else, perhaps traveling slower, or aiming more toward the ground. Perhaps it's already slammed into the barn wall.

A very fast camera can capture a bullet's precise location, but we can't tell exactly how fast the bullet was traveling by looking at the picture. So we take a second picture of the bullet at a different instant and extract information by comparing the two photos.

Think of all the measurements that are changing as the bullet travels: If we aim slightly upward, it accelerates out of the barrel, then its speed decreases as it climbs against gravity, then increases as gravity pulls it downward while air resistance slows it down. Its elevation increases then levels off, then gradually decreases as it falls to the ground. All this while, the distance it has traveled is increasing and time keeps ticking away steadily.

Whew! Time is changing, vertical distance, horizontal distance, and upward or downward angle of trajectory. Except for time, none of these is changing in a nice steady way.

Yet all these measurements are clearly related. As one changes the others change. The more time goes by, the farther the bullet will travel, for example. On the other hand, as time ticks away the bullet's elevation first increases, then decreases. Clearly the relationships are complex. Simple arithmetic can't express them adequately.

By comparing only two pictures, our information is crude. If they're a tenth of a second apart, we'll be able to calculate the bullet's average speed during that tenth of a second. But we have to guess whether it was accelerating or decelerating. We can only guess it's actual speed at any instant between the two pictures. More measurements would help, and the closer together we can take those photographs, the more accurate our average will be.

Or we could use two new tools, "functions" and "calculus."

A "function" is a mathematical procedure that follows specific rules of behavior. Perhaps we'll describe the relationship between time and distance as a function, or the relationship between the bullet's elevation and its speed as a function. We may wish to combine several functions into one expression that describes the way the variables interact in all their complexity. The advantage of describing things as a function is that mathematicians have spent a lot of time learning how functions behave and we can trust the answers they give us. Functions are not necessarily difficult, they just follow certain rules. "Doubling" a number is a function. "Adding one" to a number is a function. We'll spend more time describing functions later.

The second tool, calculus, is a set of procedures which convert any function or combination of functions into a new mathematical creature. This new creature, called a "derivative" describes the relationship between the variables at any point you want to choose The process of changing a function into a derivative is called "differentiating the function." A function is a procedure, the action to be taken, like "double the number." The derivative is the relationship between the two variables expressed as a rate. If the function says, "I will double whatever number you give me," (the process) the derivative says, "One variable is changing twice as fast as the other" (the rate).

Calculus transforms functions into derivatives. It compares the number of laps you ran with the elapsed time and tells you your average speed. Once you have a derivative, you create an algebra problem which uses it to answer the question. Perhaps the question is, "when is the function neither increasing nor decreasing?" In that case, you might create an equation which has the derivative on one side of an equals sign, and zero on the other. Algebra is the tool you will use to actually solve the problem and extract the information. If x is in the derivative, and you want to know how big x is when the derivative is zero, you merely solve the equation like any other algebra problem.

If the relationship between two variables doesn't change, algebra or geometry or trigonometry can knock it down. You don't need the big gun of calculus. If the two variables change in a way that can't be described by a function (in this context we mean a specific kind of algebraic expression) then even calculus can't help you. But if the two variables change in a predictable way that can be described as a function, calculus lets you extract lots of good information from it. It can work this magic even on an extremely complex function, or several functions combined into one massive and ugly briar patch of mathematical stuff. This is a huge shortcut and a powerful tool.

It does this by giving us a set of procedures to apply to the various kinds of functions as they appear singly and in combinations. When you apply one of these procedures to a function, you get a new function which represents the way the two variables change in relation to each other for any value. This new function is called the "derivative."

There are perhaps a dozen types of functions that are commonly used, each one has a procedure to convert it to its derivative. There are five ways that functions can be combined. Each of these combinations has a procedure. The good news is that some of these procedures are so simple you'll learn them in 10 minutes. If you only wanted to learn the useful tools of calculus, that information could be covered in a week. The bad news is that your instructor will also want to prove to you that each of these works, and why, and the logic behind them. He will want you to be able to relate the derivative to a graph. This will occupy the rest of the semester.

Calculus enables us to create a function that describes the changing relationship between variables so that we can answer interesting questions with one calculation, rather than the dozens or hundreds of calculations that would be necessary without it. For example, if a car is accelerating very slowly but steadily, we can use calculus to predict exactly where it will be 10 minutes from now, and how fast it will be going. If it's accelerating into a head wind that is gusting in a predictable way, we can still find the answer in one calculation by combining the car function and the wind function, then finding the derivative. If there's a mouse running back and forth across the front seat in a predictable way, and we want to zap it with a laser from an airplane flying fifty miles above it but with the gusting wind behind it, we can create a single algebra problem that will tell us exactly where

21

it will be fifteen and a half seconds from now accurately enough to burn our brand into that mouse's back. All in one calculation. Don't let them tell you calculus isn't useful.

But it's easy to lose sight of this usefulness when you're knee deep in finding derivatives. So much time and effort are devoted to the process of transforming functions into derivatives there is little time to actually imagine situations that would provide the original information. Without them, the process itself can seem completely abstract and removed from reality. You rarely get to do the first part of the problem, the part that relates to real life.

Compounding this is the fact that, once you've created the derivative and created the equation that would churn out an answer, you often don't get to (some students might say "have to") solve the problem. Solving algebra problems takes time and your instructor will want you to spend that time learning calculus, not solving problems. Besides that, most people survive algebra classes without ever getting very good at it. When they try to take a calculus problem all the way to an answer, they stumble on the algebra, not the calculus, and get the wrong answer. Math instructors don't have time to re-teach algebra and it wouldn't be fair to judge a student's understanding of calculus by his algebra skills. The efficient solution is to have students stop once they've completed the calculus portion. You don't get the satisfaction of a meaningful real world answer either.

So, much of your time will be spent in that middle area, the part of the process most removed from the real world. People who go on to take advanced business, engineering, or physics classes will get to use the whole process to answer interesting and useful questions. But during their calculus course they will be finding the maximum volume of various cardboard boxes and fencing in sheep they'll never own and predicting the population growth of viruses they hope they'll never encounter.

Another reason calculus seems bizarre and abstract is that it is often explained in terms of lines on a graph. Just so you're well armed, we'll do it too. You will never encounter a calculus instructor who does not think of calculus in terms of lines on a graph, because that's the way it was taught to her. But the actual process of performing calculus problems, or of using it in the real world, has absolutely nothing to do with the graphs you'll see. It's like teaching someone to dig a ditch by comparing that process to listening to music. In that ditch-digging class you'd listen to increasingly complex music that is somehow analogous to leaning into your shovel and hoisting dirt. The music will be on the test, you'll be graded according to how well you can sing along, but when you're out in the hot sun it may be hard to remember how each particular harmony applied to the dirt before you. Students with a good ear will get better grades but they won't necessarily dig better ditches. There are dozens of visual analogies that would work as well, but the world has settled on using graphs. Just remember that's all it is – a visual representation of the logic behind the math. Calculus is a process of creating functions that describe the changing relationship between variables so you can use these functions in an equation to learn information. The graphs are a way of convincing you that the relatively simple processes you'll actually use make sense. They can be useful road maps, but they won't drive the car for you.

Beyond these obvious obstacles, calculus books are written by math instructors and the courses are taught by them. Now, we love math instructors. Some of our best friends are math instructors and they are, as a rule, delightful people and good banjo players. Their one character quirk is that they revel in precision. Because they love precision, they can't stand the idea of teaching you a concept until they can prove its truth to you. Unfortunately, some simple concepts can't be formally proved until you grasp some very slippery and abstract ones so they feel they must teach you the slippery ones first. They refuse to teach you to drive until you can repair your car's brakes. They would never in a million years throw out the word "function" as we have done without spending a week explaining it to you in all its glorious precision. Be gentle with them. This precision is the love of their lives, whether you share that love or not. Don't call their baby ugly.

They also have spent their lives studying lines on a graph, and to them, that's the easiest way to explain the concepts to you. If you're a person who senses a big gap between how much cat food is in a dish and the way some squiggly line looks, you'll have to spend extra effort making that connection. They're not going to abandon several hundred years of tradition for us.

Math instructors use calculus every day in their jobs, just like engineers and astronomers do. The difference is that math instructors don't use calculus to design bridges, they use it to solve calculus problems. In their daily lives, this is what calculus is useful for. To them, solving calculus problems is a fun game, like chess, full of strategies and opportunities for cleverness. But if your chess instructor learned to play chess by drawing lines on a graph paper that somehow represent strategies for the middle of the game, and most of the chess class involves drawing graphs, and you rarely got to play, you might simply throw up your hands and go back to poker.

Perhaps the biggest distinction between calculus and other varieties of math is a subtle change in the way it *feels*. Algebra and geometry and their cousins feel somehow stationary. We walk through

24

the steps as if walking through a garden, every statue in its place, every tree planted, each pathway marked for all to see. We follow rules and proofs as if they were maps and find the predictability comforting. On the other hand, calculus feels fluid, constantly in motion, like a train moving through the garden. We approach precise answers, moving ever closer, and when we're close enough, we jump off the train. Different variables push and pull us at the same time. In the other forms of math, we look at a movie one frame at a time, but in calculus it feels like the film is rolling through the projector while we're taking our measurements. This doesn't mean it's harder. But this strange fluid feeling can be disconcerting. We're walking through a familiar garden when suddenly trees we've known for years suck their roots from the ground and start walking toward us.

Watch for this feeling. If you're unprepared it may startle you. If you never sense it, you're probably merely going through the motions and aren't really grasping calculus.

People expect the world to follow the patterns of simple math, and are uncomfortable when they don't. Shadows grow longer as the sun sinks in the sky, but they don't grow according to some nice linear function, even though we instinctively think they should. Calculus translates non linear processes like that into forms our linear minds can grasp and deal with.

Differential and Integral Calculus

Two catmobiles are driving side by side down a superhighway. A rope stretches between them. As long as the two are traveling at identical speeds, the rope does nothing. A cat in each car can hold it casually with one paw.

But if one car moves a little faster than the other, the cars will gradually move apart and the cats will have to play out more rope. If the slower car accelerates and catches the first car, the cats have to pull rope in.

Calculus is the study of the rope.

Calculus is usually taught in two distinct sections. In differential calculus you'll start with a function and convert it into a new function (called the "derivative") that expresses the relationship between the changing variables in the original. Are the cats pulling rope in, letting it out, or is it stationary? The function that describes the activity of the rope itself is called a derivative.

The derivative describes how fast the cat is letting rope out or taking rope in at any spot on the road. It is a function that describes the relationship between the two cars. You can use this new function to deduce information. You might say, for example, "what does y (the second car) equal when the derivative is zero?" That is, what's going on with the cars when the cats don't have to let out rope or pull it in? When you solve that, you'll know where the second car is when the two cars are traveling at the same speed. Or "is the derivative positive or negative when x (the first car) is 100 feet down the road?" That is, is he letting out rope or pulling it in? Or, "when does he stop letting out rope and have to start reeling it in?" That is, when do the cars stop moving apart and start moving closer together?

If you're in the real world rather than in math class, you won't start with a function. You'll start by taking measurements of the cars. To get useful information, you need to measure the location of each car at two different instants. Between your two measurements, perhaps car x moved 10 feet, perhaps car y moved 15 feet. This actual change in measurements is designated by the Greek letter "delta" which looks like a little triangle in front of each number. The smaller the distance between your measurements the more precise your results will be. Think of delta as translating to "an increment" or a "tiny change." Because we get more accurate answers with tiny increments, it's useful to assume that we're talking about little bitty changes when we see a delta. The relationship of the movements is Δy over Δx and looks like this:

$$\frac{\Delta y}{\Delta x}$$

That is, y changes some specific tiny amount every time x changes by some specific tiny amount. The deltas remind you that you're dealing with actual real world measurements, not math. Once you decide to use math upon these measurements, you'll replace each delta with a small letter d. This means you are moving from the real world into the world of math, where different rules apply. Now the same relationship looks like:

$$\frac{dy}{dx}$$

The subtle difference is that we are no longer thinking about actual measurements, but rather about the relationship between the two measurements. When you saw deltas, you were talking about cat-mobiles moving a specific distance. If y moved 10 feet, perhaps x moved 15 feet. That was the actual measured difference. This new thing, with the small ds, is no longer specific measurements but a mathematical relationship that we can manipulate. We call these new creatures "differentials." If we have that information, we can replace dx and dy with those numbers. If Δy is 10 and Δx is 15, we are now discussing the ratio of 10 to 15, which is the same as the ratio of 2 to 3. We could substitute 2/3 for 10/15 and come up with the same answer in a math problem. That's much different than saying one car moved only two feet and the other three feet, when we know they each traveled five times that far. Instead it says that for every three foot movement of x car, the y car moved 2 feet.

Of course, we could also measure our cat's rope activity during that brief portion of the drive. He might be pulling rope in at some rate, or letting it out, or chewing on his tail. This information isn't very useful if the measurements are several minutes apart, however. He might have done any number of things with that rope, and chewed on his tail several times during a minute. So, we want our measurements just as close together as possible.

Differential calculus gives us tricks to tell what the cat is doing with the rope. All our answers will involve a relationship. If location is one variable and time is the other, we express the relationship between the rates of change of each as speed. You can figure out many bits of information if you know a car's speed. If revenue is one variable and number of units sold is the other, the derivative is called "marginal revenue."

$$\frac{\Delta r}{\Delta u}$$

MARGINAL REVENUE

Integral calculus is the reverse process of differential calculus, just as division is the reverse of multiplication. You might know what the rope-cat is up to, but now you want to know how far the cars traveled between measurements. You might know the shape of the vase, and how fast you're pouring water in, but now you want to know how high the water level is at some instant, or how much you'll pour in during some interval of time. In integral calculus, you'll start with known relationships and your answer will be specific quantities.

The Difference between Algebra and Calculus

Algebra is a set of tools to *solve* problems.
Calculus is a set of tools to *create* problems. Huh?

That is, calculus allows you to express relationships between dynamic variables so that you can create a meaningful algebra problem from the information at hand.

Another way to think of the difference is to consider a chart like this:

time	x	y	function
1	2	4	$y=2x$
2	4	8	$y=2x$
3	6	12	$y=2x$
4	???	16	$y=2x$

Algebra is concerned with identifying information that's missing from the chart. It asks the question "what is x when y is 16?"

Calculus is interested in something that is not on the chart at all: how much does y change every time x changes a certain amount?

$$\frac{16 = 2x}{2} \qquad x = 8$$

This new relationship, the derivative, is not data. It's nothing we measured, it's the relationship between the columns. If the information is simple, if the relationship isn't complex, the derivative isn't going to be especially interesting:

time	x	y	function	derivative
1	2	4	$y=2x$	$y\,'=2$
2	4	8	$y=2x$	$y\,'=2$
3	6	12	$y=2x$	$y\,'=2$
4	???	16	$y=2x$	$y\,'=2$

Calculus asks "how is the relationship between x and y changing?" The answer is the derivative which tells us that y changes twice as fast as x, and the relationship doesn't change no matter what x or y or time you're interested in. That is,

$$\frac{dy}{dx}=2 \; \approx \; slope$$

tells us that, for every incremental change in x, the change in y is twice as big. But this is trivial and obvious and boring. Calculus only becomes interesting when you're trying to solve problems that are too difficult for lesser forms of math. Unfortunately, even the simplest of the interesting problems look as frightening and foreign as Halloween in New Orleans to most of us. On the other hand, if you look at simple, easy to understand problems from a calculus perspective, the process will bore your math instructor. We all agree that it's better for you to be frightened than for your math instructor to be bored. In class, you won't spend much time on simple, easy to understand problems.

The Power of Calculus

Differential calculus is used three ways, each quite different from the other. Mathematicians and cats instinctively understand this, so they tend to gloss over the distinctions, which confuses the rest of us. The distinctions arise from what question we want to answer.

1. *Precision*: Calculus helps us find values at a specific instant. If we want to know the relationship between two variables at a specific point, we use calculus one way. If the variables are time and distance, the relationship between the two is speed. If we want to know how fast a bullet is traveling after exactly one second, calculus gives us a more precise answer than we could get by any physical measurement. How can this be? Once we have described the relationship in terms of a function, we can dissect that function into microscopic bits because it's just math. How fast will the bullet be traveling one millionth of a second later? Easy to find with calculus, but impossible with a stop watch and tape measure. In these situations, the power of calculus is its precision.

2. *Approximations*: Sometimes we want to know how extremely small changes in one variable affects the other. For example, if we gradually increase power to one engine of our space craft by one percent per year so that its path is no longer perfectly straight, and it travels near the speed of light, where will it be in a hundred years? In this situation, we don't use calculus because it's precise, we use it because it's quick and easy and close enough. In fact, we know right from the beginning that our answer will be slightly wrong, but we'd rather get an extraordinarily close approximation with one quick calculation than spend two days computing. Remarkably, as the numbers

get larger, our error gets smaller. If we are considering the weight of a blob of bacteria as it grows, the error may be measurable when the colony is microscopic. By the time the blob fills our living room, the mathematical imprecision will be so small as to be insignificant. In this type situation, we do not use calculus because it's more precise than measuring, we use it because it's by far the easiest way to get a great approximation of the answer. This is different from our first example.

3. *Minimums and maximums:* The third powerful way to use calculus is to predict which combination of variables creates maximum or minimum results. When will the bullet reach its highest point? How do I make the most profit? When will the bridge break? With calculus we can answer each question with a single calculation, rather than dozens of attempts at trial and error.

The processes used in each type of problem are similar, so it's easy to glide along through a calculus course, content and cat-like, solving problems and getting good grades, without ever understanding how or why you might actually use it. In that respect, having read this page, you now understand the power of calculus better than many people who have actually studied it. But, of course, not as well as most cats.

How It Works

The calculus machine only works on functions, so the first step is to translate real-world information into functions. A function is just an abbreviation for some specific mathematical process you'll subject your unknowns to. Calculus becomes useful when two or more non-steady processes are going on at the same time, each one

NON. STEADY STATES

having a different effect on the outcome. A rocket's engine pushes it upward in opposition to gravity. Without gravity its speed would accelerate upward as it overcame inertia. On the other hand, without the thrust of the rocket engine, gravity would pull it down. Plus, the weight of the rocket itself is changing as it burns fuel. Some forces are pushing it upward, some pulling it down, and all are acting at the same time. Each one of these forces would be described as a function, then all of the functions combined into one big rat's nest.

Luckily, there are only a few forms this rat's nest can actually take. A couple of very smart cats named Newton and Leibniz spent many fun hours figuring out how any group of functions could be manipulated to represent the slope of any point on the line they would graph. The process they used, while logical in retrospect, involves a lot of algebraic mumbo-jumbo. Your calculus instructor will feel that reproducing Leibniz's thought process builds character. By the time he's done with you, you'll have enough character to build a whole new person. You'll never use those calculations again.

The end result, however, is a handful of rules or processes which will astound you with their simplicity. They are the constant rule, constant multiple rule, sum rule, power rule, product rule, the quotient rule, and the chain rule. You could write all of them on one three-by-five card. Depending on the form of your rat's nest, you will choose one of these little processes to transform it into a derivative. The derivative is a new expression (which will look like any other plain old expression) that describes the slope at any point on the line the information graphs. It describes the rates of change of the variables as they interact and is true at every point on the line.

Finding the derivatives of functions is the central task of calculus. There's one similarity to algebra that seems so obvious to the folks who write Required Calculus Textbooks that they may neglect to mention or emphasize it. In algebra, you could maintain the equality of an equation by "doing the same thing to both sides of the equation." If you add 10 to each side, you've changed the equation, but it remains true. If x equals z then $10 + x$ also equals $10 + z$. Similarly, you can take the derivative of both sides of an equation without affecting the equality. If *raccoon = tractor*, then the derivative of *raccoon* also equals the derivative of *tractor*.

Tangents

The Cat Home Planet is not a sphere. It's a huge, flat pancake shape suspended in space like a galactic compact disc. The headlights from a catmobile on that planet never touch the ground. The perfectly horizontal light beams stay a foot or so above the ground, parallel to it in their laser way, until they shoot off into space.

On Earth, if you park your car on a big flat parking lot, the car and the line of its laser-light-beam-headlights are also horizontal. They never touch the pavement, but shoot off a foot or so above it.

You already know the word horizontal. In general slang we use it to mean "parallel to the ground," or "parallel to the horizon." The surface of a pond is horizontal. We think of it as a straight line, just like our laser lights. It's the opposite of vertical.

But Earth is not flat, it's a sphere. It only seems flat because it curves on a scale too large for our daily perception. The ocean is not flat, and neither is the most massive and level parking lot we can imagine on Earth. On a planetary scale, the word "horizontal" needs a little help.

If our catmobile is sitting on a raft in the middle of the Pacific Ocean on a calm night, its head- and taillights will shoot off a foot or so above the water, absolutely straight. Gradually the Earth will curve away from the light beam while the laser continues in a straight path. Finally the laser beams will shoot off into space. Instead of calling the straight line of our head and tail lights "horizontal" we call it "tangent" to the surface of the Earth.

Just like the word horizontal, the most important feature of the word tangent is the direction of the line. A tangent line touches a

35

curve at one point like a yardstick touching a volleyball. Of course, if the curve squiggles around, the tangent line to one point may cross it or touch it elsewhere as well. In calculus, we'll be finding tangents of football shaped curves, and potato-chip bowl curves and frozen earthworm curves. The reason we do this is because we want to translate information into math. To make it easy to visualize, we use this math to create graphs. The graphs can be as simple as a curved line, or they may look like a loose extension cord thrown onto the garage floor.

Just as the angle of the catmobile's headlights represents the relationship between the height of the front tires and the height of the rear tires, the tangent line at any spot on a graph represents the relationship between the two changing variables at that point. After that point, the graph curves away and the relationship changes. The length of the tangent line isn't interesting to us; what's interesting is its steepness (or slope). This steepness expresses the rate of change between two variables at a particular point on the graph of a function.

Graphs

As the cats like to say, graphing things is so easy even a dog can do it. It's a way of visually comparing two changing quantities or two "variables." Often, one of these variables is time. Usually the other one is cat food. After dinner on Saturday we have 10 cans of food; if we eat one every day, the graph will look like this after five days:

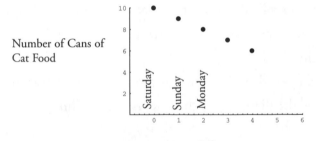

36

The horizontal line is the x line or "x axis." (*axis* is Latin for "line.")

The vertical line is the y line or "y axis."

We can describe any spot on the graph by giving it "coordinates." That is, we tell where it is compared to the y axis and the x axis. The x coordinate tells how far right or left the spot is, the y coordinate tells how far up or down it is. For consistency and to be alphabetically correct, we always say the x coordinate first. Before we plug any information from the real world into our graph, we describe any point as (x,y). Yes, we put them inside parentheses just like that. Yes, (x,y) is a generic representation of any point on any graph. As we learn information we replace the appropriate letter with the information. If all we know is that it's Thursday, we might change it to (Thursday, y). Since we don't yet know how many cans of food there are we leave the generic y in place until we do. In this case it's easy to see what y should be. We extend a vertical line up from Thursday and see how high it is when it crosses the line. It's the same height as 5 cans. On Thursday we'll have 5 cans of food.

This works because we can see the "slope" or slant of the line. Slope means the steepness. Slope represents the rate of change of both variables. For each day that passes, our supply of food is reduced by one can. Slope is always expressed as something that looks like a fraction. Again, for consistency, it's described as the "rise over the run." That is, the number representing the vertical variable (y) is on top with the horizontal variable beneath it. It looks like this

$$\frac{y}{x}$$

or in our example:

$$\frac{-1 \text{ can}}{1 \text{ day}} = 1 \text{ can per day}$$

The slope of a straight line is predictable. Once you know any two points on it, you can figure the slope of any other part of it. Some lines are less obvious because the variables that we're graphing keep changing, so the line curves. But the same principle applies. If we can figure the slope of the curve at the spot we're interested in, we can probably deduce other information from that.

To solve many problems, we can either make an actual graph, or we can express the relationship between the variables as a ratio and use it in an equation.

If we have enough information, we can graph two points, connect them with a straight line, and determine its slope. But the rules of the game tell us we can only use a straight line to connect them.

Many things, alas, don't change in such a regular way. As squirrels drop from trees their speed accelerates. As a herd of mice reproduces and the babies mature and start having babies, a straight line is no longer adequate to describe their population. Such important information demands a refinement of our math.

Calculus is a way of determining the slope of a *curved* line at any given point. If a catmobile parked on that spot, calculus lets us describe the direction its headlights point. Then we use algebra on this headlight information to find the answer to some feline question.

The calculus process isn't worth the effort in simple problems like our cat food example. If you've never stared nervously and lovingly at the last can of cat food, knowing shopping day is still several days away, it may seem trivial to you. It merely illustrates the logical thinking that calculus extends into much less intuitive arenas.

The slope of a line represents the rate of change of the two variables. A fast rate of change will be steeper than a slow rate. If you describe it as the rise over the run, the difference between the top number and bottom number will be larger. 14/1 is a lot steeper than 6/1. For every step you move right (the run) your line will line will go up 14 steps, (the rise) compared to 6 steps.

In calculus, the rate of change (the slope) is called the "derivative." When we're trying to find the derivative of a line, we're trying to find its slope, and we call the process differentiating. We're answering the question: how much does the rise change compared to the run in this tiny section of graphed line?

If you're actually counting mice or timing the fall of squirrels, you represent the amount of change between two measurements with a little triangle. This is the Greek letter for "D" and it's called a delta. If you see "Δx" that means "the change in x" from one measurement to the next. The delta sign lets us know this is information from the real world.

The slope of the line between two points depends on the amount of change in x and the amount of change in y. You could write it like this:

$$\frac{\Delta y}{\Delta x}$$

If you're converting the information into mathematics you'll use the small letter d to represent the change from the first measurement to the second. dx means "the amount of change in the variable x."

The derivative will look like this:

$$\frac{dy}{dx}$$

If our information forces us to graph a curved line, the distance between the two points we measure becomes important. A straight line between two points on the curve won't represent the slope at either point exactly. Two points that are far apart will give us only the crudest idea of the slope of that part of the curve. The closer together the points, the more accurately we can describe the slope.

But we can't let the two points get so close together that they're really the same point. We need two different points to determine a slope. This creates a little dilemma.

Ideally we want them so close together we can't even measure the space between them, but if they're that close, obviously, we can't measure them. We never let the distance actually become zero, but we can let it become such a tiny number as to be insignificant to our math.

Calculus can be very accurate at predicting the results of variables which are constantly changing. It is most accurate when the section of the line it considers is extremely small. The change represented by dx or dy is going to be tiny most of the time.

Perhaps we've come to think of calculus in terms of graphs because math is well suited to solving problems that involve straight lines, but we don't live in a straight line world. Geometry gives us lots of rules (some would say "way too many rules") to deal with straight lines and the shapes we make by combining them. Trigonometry provides a whole feast of things to do with triangles, although some of us will never be that hungry. Algebra easily handles linear equation, the ones that graph straight lines. Straight-line math seems intuitively right, even to students with no particular interest in math.

Once equations start graphing lines that curve, math feels a lot less comfortable. Even a simple circle conjures up weird things like *pi*, a number that goes on forever, and numbers with evil-looking exponents. Complex curves seem dangerous.

Calculus takes the curves graphed by exotic functions and allows us to think of them as tiny straight lines stuck together. It does this by considering one tiny portion of the curve at a time, the smaller the better. We are always asking the question, "If *x* changes a certain amount, how much does *y* change?" If the equation graphs a curve, and we use little straight lines, we're always going to be just a tad off. But if your measurements are close enough together, it's accurate enough. It's like asking, "how flat is the Pacific Ocean?" The true answer is "not flat at all." It curves with the earth's surface. But if your task is skipping stones across its surface on a perfectly calm day, you don't need to take measurements a thousand miles apart. Two measurements fifty feet apart connected by a straight line will give you an approximation of the shape of its surface that is more accurate than your throwing skills.

Functions

The whole idea of math is to translate information from the real world into numbers and symbols we can manipulate to predict things. Each branch of math translates information a little differently.

Arithmetic uses numbers. We convert a pile of apples into a certain number of apples then manipulate this number using the rules of addition, subtraction, multiplication and division.

Algebra primarily uses polynomials. We express the things we don't know in terms of the things we do and wind up with bundles that contain both numbers and unknowns. We call these bundles polynomials, and algebra is the process of manipulating them.

Calculus uses more general functions. A function is a mathematical process we perform on a number. It can be a simple process, like "double the number," or it can be more complex, like "square a number, then double the result, then subtract the speed of light from it." Some processes lead to more than one correct answer. Years ago they categorized functions according to how many correct answers they could generate. There were "one valued functions," "two valued functions," etc. That is, if you applied the procedure to a certain number, how many correct answers would it yield? Today we only consider "one valued" functions to be true functions. Other math procedures are not functions and don't get to play with us right now.

Cats think of functions as precise little dances. If they begin the dance at one location and perform each step correctly they arrive at some particular spot. If you know where they start, you'll know where they end. It also works in reverse: if you know where they end up, you know where they started.

42

One cat, let's call him "Garfield" (after the president), always does the dance called the watusi. The dance is very precise: he takes four steps forward, two steps to the side, eats a snack, then manages one more step forward, then settles down for a long and well-deserved nap. The dance is made up of all these steps. Simply shortening it to "watusi" helps a bit. But in math, we can't possibly name each and every combination of procedures. So we give them the general name of function and then describe them in more detail for each problem.

If you see *watusi* (*x*) in a problem, you know what dance Garfield does and that he starts at *x*. If you know where to start, you'll know where he ends up.

watusi(starting spot) = ending spot

As you begin studying functions they will usually use the letter *f* to represent an entire function. As you encounter situations involving more than one function, they will most often use *g* or *h* to symbolize different functions. If something is a function of time, it's not uncommon to use *t* to represent time.

Different cats do different dances so we need to label them appropriately. We don't want Garfield break-dancing when he's supposed to be waltzing. When we write down a function we'll always have two elements: the abbreviation for the procedure, and then next to it (in parentheses) the name of the starting spot.
When we see

$$f(x)$$

we know that *f* is the abbreviation for the procedure and *x* is the character that will have that procedure done to it.

Here's a great opportunity for confusion. You have always thought that when you saw something in parenthesis you were going to be multiplying. Those days are over. If you see an f before parentheses, they are NOT telling you to multiply f times the contents of the parentheses. They're telling you that you will perform the f function upon whatever is inside the parentheses. In the same problem we might see

$$g(t)$$

and we know that g represents a procedure we're going to do to t as soon as we know what number it represents.

Here's another fun way to confuse you: the first time you see $f(x)$ in a problem, they are probably defining the function for you. They will say $f(x)$ equals something, and that's how they let you know what to substitute.

$$f(x) = 2x + 34 - x^2$$

This is not telling you to do anything, it's just their way of giving you information about the f dance in this problem. They have said, in their very efficient math language: "This is what the function we're going to call f does: it doubles x, adds 34 to the answer, then squares x and subtracts the result from answer above." (Note: A math-lover will panic or become indignant that we've put these in a non-standard order. If there's an x^2 it should come first, then $2x$, then 34. But for illustrating this idea, it doesn't really matter, does it? This is just a test to determine how entrenched you are in math conventions. Now you may return to your reading.) Now wherever we see $f(x)$ in the problem we can substitute $(2x + 34 - x^2)$ for it.

If we discover that x equals 63, we can substitute 63 for the x's by saying

$$f(x)=2x + 34 - x^2$$

or

$$f(63) = 2(63) + 34 - 63^2$$

Functions resemble a recipe that doesn't tell you what ingredients to use but gives specific instructions as to what do with them.

If the function of x is to double it, we would show that as:

$$f(x) = 2(x)$$

Now, anytime we see $f(x)$ we know what we're supposed to do. As soon as we know what x is, we replace the x with the real number and multiply it times two.

In a graph, the vertical location of a point is dependent on the horizontal location you're interested in. You could say y is a function of x and write it like this:

INDEP

DEP $y = f(x)$.

Most of the time we're going to be looking for y, which will be a function of x. But if you know the vertical point, you might be looking for x. In any process you could consider either x or y the dependent variable. It's determined by which one you know and which one you're looking for. The one you know is the independent variable. That's where you start your dance. After performing all the steps, you'll arrive at the dependent variable. In math classes, y will almost always depend on x, and x is the horizontal axis of the graph.

In our cat food example the number of cans (y) is a function of the number of days since Saturday (x); y is the dependent variable.

$$\frac{-1 \text{ can}}{1 \text{ day}} = 1 \text{ can per day}$$

What process can we describe that will always give us the correct number of cans? This is the first challenge of calculus. We need to convert situations into functions. Saturday is day zero. Sunday is day one, Monday day two, etc. By examining the information we've graphed, we see that if

$$f(x) = 10 - x$$

this function is true every day for the next week or so. That is

$$y = f(x) = 10 - x.$$

Once we have the problem expressed as a function, or a series of functions, we use one of the rules of calculus to convert it into a derivative. Then we use that derivative in an equation which we design to answer the question we're interested in. Since the derivative will just be a bunch of numbers and unknowns, you'll solve it exactly as you would any other algebra problem.

Why in the world would you bother? Well, you wouldn't in this case. The problem's too simple to be worth the effort. The reason it's simple is that the function describes a rate of change that's constant, it graphs a straight line. A math guy would know this in an instant just by looking at it because the equation doesn't contain any exponents or other nonlinear functions built into it. It's called a linear function because it graphs a straight line. If you know the slope of one

46

section, you know the slope of any section. But when the rates of change are not constant you're in for many hours of calculating if you don't know calculus. The functions that describe them will have unknowns squared or raised to other powers, or other nonlinear forms, like trig functions or natural logarithms. The lines they graph won't be straight.

You don't need to get into astrophysics to run into changing variables.

If you're filling a rectangular aquarium with water, the water level is a function of time and it's a nice linear function. You can predict where the water line will be in a minute, or three minutes, or 10 minutes.

If you're filling a spherical aquarium, the water line is still a function of time, but it won't move at a nice steady speed. The function that describes it will have at least a squared variable in it, the line it graphs will be curved, and you'll be tickled pink you learned calculus so you can predict where it will be at any second.

There are several types of functions that occur so often they have their own little abbreviations. The six trig functions, for example are usually represented by their nicknames. These are functions imported from trigonometry to help describe angles and rotation in a way we can use in calculus. If you see sin(x), that's what's going on.

Linear Mice

There's a mouse in your living room and you want to describe its location to a cat. One way you could do this is by saying it's three feet east of the cat food dish and ten feet north of it. With this simple system, popularized by Renee Descartes, (and named "the Cartesian system" after him) you can pinpoint that mouse to any cat.

If the mouse starts jogging in a straight line, at a steady speed, you can use the same system, but you have to figure in how long it jogs. Each second it moves it will be a slightly different distance north of the bowl and a slightly different distance east of the bowl. Perhaps each second it travels one foot east and one foot north. After only a couple of measurements you will recognize the pattern. You make a little chart.

t	x	y
0	3	10
1	4	11
2	5	12
3	6	13

If you aren't too distracted by the sight of a mouse jogging across your floor, you can probably instinctively continue the chart and predict where the mouse will be in seven seconds. You can also predict whether it will run into the north wall first, or the east wall, (which is eight feet east of the cat food bowl). You see the pattern in the chart, so you know how to extend it. Sometimes the pattern is much less obvious. Luckily, there are a couple of other ways to extract the same information.

To get the same information, we could express the chart as an equation:

$$y = x + 7$$

then substitute the information you're curious about and solve the equation.

How far north will the mouse be when he's 7 feet east?

$$y = (7) + 7$$

Or you could make a graph from the information on the chart. Because the information is graphing a straight line, you take your ruler, extend the line and extract the same information.

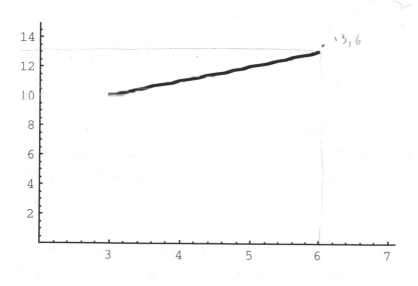

$$f(x) = x + 7$$

The steepness of the line is called its "slope." It's a visual representation of the relationship between two changing measurements. The relationship doesn't change in this example because our furry friend moves in a straight line. That is to say, it's a "linear mouse." The chart of a linear mouse is easy to extend, it will graph a straight line, and the equation that expresses is will be linear (that is, it will contain no exponents or other non linear functions).

If you tell me a linear mouse's location at any instant, and I know the slope, I can tell you precisely where it will be one second later, or a tenth of a second later.

In calculus the slope is called the derivative. We describe it just like we did with the mouse: either as a formula or as a line on a graph. Whether we represent it as a formula or a graph, the slope (or derivative) is the relationship between two changing variables.

The first huge leap from the real world to the abstract world that cats love so well happens right here. And at this precise instant many people fall off the train that's bound for Calculus Enlightenment without even realizing it. Traditional calculus courses often barrel right past, unaware that half their passengers have suddenly disappeared. It's not a big difficult concept, it's a tiny, subtle concept that is so easy your instructor may not think to mention it:

Up to now, our graphs and equations are representations of real information, of data we can list on a chart. In algebra or geometry every x and y, every a, b, or c stands for something in the real world. But the slope isn't data, it isn't information from a chart, in fact it's not from the real world at all. It's just a mathematical way of expressing the relationship between things that we can chart. What's interesting and subtle is that we now begin to treat the slope, or derivative, exactly the way we treat any number or expression. We might have fifteen cats in our equation, and eighty pounds of cat food, and any number of wool sweaters, and then this other thing, this derivative that has no odor or weight or taste or usefulness whatsoever in the

50

real world. It only exists in math, and yet we treat it with as much respect as a nice bowl of liver.

It creates confusion because it looks and acts like any other number or expression or function, but it arrived by magic. It's a very pleasant ghost singing in our choir. It sings the part just fine, it doesn't bother any one, and yet it makes us nervous and we'd just as soon not stand too near it. Mathematicians are used to ghosts in their choirs, and they don't even notice it. But you understand that this is not like the choirs you have sung with in the past. Every singer in those choirs represented something you could measure in the real world. Every x was a certain number of apples, every c was a line in a triangle. The fact that the instructor is completely oblivious to this new apparition only makes you more nervous.

A derivative is like the changing nature of the friendship of two people: It's absurd to imagine the friendship existing and changing without either of the people, yet in calculus that's just what we do. We call the friendship itself the derivative and use it in an equation without using either of the two people. The derivative is the changing harmony of two musical parts. Where is the harmony without either part? It certainly can't be heard, in fact it doesn't exist, but it could be mapped mathematically. A derivative would describe the harmony itself rather than either of the two notes that create it. We might say, "the harmony part will be an octave higher than the melody," or "sing parallel thirds." A musician would understand this silent music, this harmony without specific notes, just as a math instructor will understand the slope of a graph or derivative of an expression. To the rest of us, this is a weird concept and very abstract. Not surprisingly, cats don't find it odd at all.

The slope is the change in vertical movement for each unit change in horizontal movement on a graph. For each distance a spider moves east (or parallel to the x axis) how far does it move north (parallel to the y axis)? If it moves north three feet ($\Delta y=3$) each time it moves one foot east ($\Delta x=1$) the ratio is obviously three to one. Whether it moves for one second or ten seconds, it will move three times as far north as it does east. This is written just like a fraction, with the y on top. Remember this by the phrase "rise over run." Our example will look like this:

$$\frac{3}{1} \qquad \frac{y}{x}$$

You can manipulate this just like you would any fraction or ratio or division problem. The ghost has slipped into the back row of the choir. In algebra it's common to abbreviate the slope with a small letter *m*:

$$m = \frac{3}{1}$$

One thing you could do with that is simplify it:

$$m = 3$$

Now we have a simple number 3 that looks like any other garden variety number 3. It acts just like a three, and will sing the same part in the choir as every other three. But it's a ghost. It doesn't represent any data on your chart, no spot on the carpet where you observed the spider. It represents the relationship between the spider's movement north and its movement east.

Just so you know that we're in a calculus class rather than an algebra class, now we call the slope the derivative. Instead of using *m*, we abbreviate "derivative of" with a little apostrophe sign like this: '

If we describe the movement of the spider as a function of *x*, then at this point the derivative is 3 and we express it like this:

$$f'(x) = \frac{3}{1}$$

Now we call the slope the derivative but it's still 3,

$$f'(x) = 3$$

You've probably noticed that algebra is generally concerned with identifying information that's missing from our chart of data:

Age	Red Cats	Black Cats
1 month	1 pound	1 pound
2 months	1.3 pounds	1.69 pounds
3 months	??? pounds	3.24 pounds

Calculus is concerned with describing the relationship between the columns. The process of describing that relationship (finding the derivative) is what one does in calculus. Perhaps you are observing the growth of both red cats and black cats. Because you hope the results of your study will lead to a Nobel Prize, you take careful measurements at regular intervals, and these are your results:

Age	Red Cats	Black Cats
1 month	1 lbs	1 lbs
2 months	1.5lbs	2.25 lbs
3 months	2 lbs	4 lbs
4 months	2.5 lbs	6.25 lbs

A calculus problem might ask a question like: "For each pound a red cat gains, how much does a black cat gain?" That is, what is the relationship between the two rates of growth? After you've done some work, you notice something startling: For cats of the same age, a black cat's weight is always equal to a red cat's weight squared.

Obviously, the red cats are house cats, and the black cats are leopards. Red cats gain weight linearly. In each month they gain a half pound. We won't need calculus to predict their weight, or their rate of weight gain. But black cats start out slower and then grow more and more from month to month. If we want to know the rate the black cat gains weight, we need calculus and it is the derivative that will tell us that information.

Sometimes, especially in a calculus course, that's all you want. The instructor will tell you to find the derivative of some function or combination of functions, you'll do it and be done. In real life you'll usually use the derivative in an algebra problem which you create to identify information missing from the chart. You might ask questions like:

"If a one year old red cat is 3 pounds, how fast is a one year old black cat growing?" or "Exactly how old is a black cat that is gaining a pound a month?" or "Exactly how old is a black cat that is gaining weight at the same rate as the red cat?"

Math is like solving mysteries. The first step is to ask the right question, to transform the mystery into math. When two or more variables are changing in a related way, it's often impossible to create a problem that represents the rate of change using only algebra. You can't ask the right question, so you can't get the right answer. Calculus is a machine for creating these kinds of problems, algebra is the machine for solving them.

Quadratic Mice

If your house became infested with mere linear mice, which always run in a straight line at a steady pace, you would need neither cats nor calculus. A fast Golden Retriever could run them down, assuming you could explain the concept to it. Unfortunately, mice come in several more complicated varieties.

If you're charting a mouse's path and your chart looks like this:

x	y
0	0
1	1
2	4
3	9

you can tell your Golden Retriever to go back to sleep – you've got quadratic mice. Quadratic mice never run at a steady speed; they are always either accelerating or decelerating. If you chart their position, you'll get a curved line called a parabola. If you transform your measurements into a function, something will be squared.

$$f(x) = x^2$$

Quadratic functions are child's play for calculus. With one process so simple you can do it in your head (the power rule) you can transform any quadratic function into its derivative. Then you ask whatever question you yearn to have answered, like, "What direction will it be heading in five seconds?" You create the algebra problem that uses the derivative to answer the question, and your mouse problem is over.

Limits

Like many other geniuses, cats sometimes create games to make simple tasks less boring. Just as basketball players find unique ways to slam a ball through a hoop to break up the monotony of earning millions of dollars for running around for an hour in baggy shorts, so cats seek out more interesting ways to solve easy math problems.

For example, in restaurants they like to tip their server a standard fifteen percent of the bill, but they refuse to stoop to simple multiplication. If the bill is $100, obviously the tip will be $15. If the bill is $83.64, the tip is $12.55, and any self-respecting cat could calculate it mentally in a fraction of a second, just as you did.

But what fun is that? To make it more interesting, cats turn it into a game. The only rule in the "figure out the tip" game is that you can never multiply fifteen percent times the exact amount of the bill. But you can multiply as many other numbers as they want.

If the bill is $123, cats might start by calculating the tip on $113 and on $133. Right away they know approximately what the tip will be. If they want to tip a little too much they leave their server $19.95. If they want to under-tip a little, they leave $16.95.

For a more precise estimate the cats will choose numbers closer to the actual bill. The tip on $128 would be $19.20, the tip on $117 would be $17.55 They can perform as many calculations as they like, so, of course, they'll do a whole bunch. Perhaps they create a little chart to keep track of their calculations:

Estimated Bill	Tip (15% of Bill)
$133	$19.95
128	19.20
123	???
117	17.55
113	16.95

56

Simple common sense tells us that the closer our estimate is to the actual bill the better our guess as to the proper tip. If we use fractions of dollars, we'll get even closer

Estimated Bill	Tip
$123.50	$18.525
123.25	18.4875
123.10	18.465
123.05	18.475
123.02	18.453
123	???
122.98	18.447
122.95	18.4425
122.90	18.435
122.75	18.4125
122.50	18.375

Don't use interpolation

At some point they are calculating fractions that are less than a penny and therefore not much use for the purpose of leaving a tip. At that point they decide to stop the game, put some money on the table and go home.

This process of deducing an answer by using increasingly accurate approximations is the "limit" process. The limit is the answer, or, in this case, the correct tip. As the estimated price of dinner gets closer and closer to $123, the limit is $18.45. In a math book that would look like:

$$\lim_{\text{estimate} \to \$123}(\text{estimate})(.15) = \text{real tip}$$

We probably want to make this look more calculusy by using function signs. We know our function is to take fifteen percent of various dinner bills which we call B, for Bill

$$f(B) = .15(B)$$

57

So now we can express the whole game as

$$\lim_{B \to \$123} f(B) = \$18.45$$

Often, we can't take direct aim at a specific number but we can create a series of increasingly accurate approximations. In these situations, the concept of the limit is useful.

The most obvious example is determining the slope of a curved line at any specific point. To measure the slope we need two points. But if the line curves, the slope won't be identical at both points. All we can really do is draw a straight line between the them and measure the slope of that straight line. The closer the two points are to each other the better our estimate. But we can't ever directly measure the slope at a single point. What we can say is that "the slope gets closer to some value as the points get closer together." Sooner or later, just as in the tip game, we're dealing with fractions of pennies and we say "close enough."

The closer together our points are, the closer the straight line between them approaches the angle of the line that's tangent to the curve at the spot we're interested in. The steepness or slope of that angle is the derivative. This explanation requires you to think of calculus in terms of a graph. But you can also think of calculus as pure math, without the graph, if you're willing to risk being burned at the stake by various math organizations. If you do, you'll discover the same dilemma that led to limits from a different perspective.

Whatever the slope is, we'll need to use it in an algebra problem to get useful information from it. Slope is always expressed as a fraction, the "rise over the run." But the Prime Directive of algebra is reversibility. Fractions that aren't reversible simply aren't allowed to

58

play. This means you can't divide by zero because you can't reverse the process and get back where you started. So at all costs we have to avoid a slope with a zero in the denominator. If we don't, we'll get something we can't use in algebra to actually solve the problem.

But zero in the numerator is okay. In fact, zero is the most ✓ interesting number in the community of slope numbers. A zero slope means a horizontal line, which means our catmobile has just crested a hill or bottomed out in a valley.

Isaac Newton tried to get around the problem of zero in the denominator by creating a whole new breed of numbers, the "infinitesimals." These were shadowy creatures of his imagination, smaller than any real number but bigger than zero. Back then, any sort of calculation was called a calculus. He called his new game "the infinitesimal calculus." Gradually this was shortened to "the calculus" and many people still call it that. It has a nice pretentious ring to it. "Oh yes, I'm studying the calculus."

It turned out that doing algebra with infinitesimals was tricky and sometimes led to absurd results. This tarnished their tiny little reputations. For some reason, Germans became especially disenchanted with them. A fellow named Weierstrass and his buddies gave us the current definition of limit. In the 1960's a logician named Robinson created a consistent model of infinitesimals and restored them to respectability. Unfortunately, logicians are themselves not respectable within the calculus community, so that project was doomed from the beginning.

The idea of the limit process is central to calculus. The limit process allows us to avoid illegal fractions with a zero in the denominator without forcing us to believe in Newton's ghosts.

Just when you think this limit business is trivial, you'll move into integral calculus and, without warning, they'll start using the word differently. You'll no longer be trying to find the slope of a point on a graph. Now you'll be trying to determine the area of the space beneath a line on the graph. In integral calculus, the word limit describes the boundaries of the area we're looking at. If you want to know how far you travel between noon and three o'clock, the limits of your integral calculus problem will be noon and three. Your answer will not be a ratio or slope, it will be a specific real quantity, like "eight miles."

Obviously, cats are very proud of the word "limit." It has baffled calculus students for hundreds of years and helped maintain the subject's mysteriousness.

Notation of Limits

The game of estimating tips by approximation hardly seems worthy of college courses and research grants. To make it more respectable, it is described with its own peculiar notation, completely indecipherable to the untrained. Like the concept, it's not really complicated, just odd. And, of course, it would be useless if the folks who grant tenure and buy computers for the instructors actually understood it, so most calculus books are careful to gloss over it rather quickly. We'll explain it, because that's what we think our job is. Just don't show this book to any college administrators or high school principals. Our entire industry depends upon their fear and awe of calculus.

We use an arrow to replace the word "approaches." So $x \to 8$ translates into "as x approaches 8."

60

We abbreviate the word "limit" with "lim." To describe the whole limit process we write the variable-arrow-limit under the "lim":

$$\lim_{x \to 100}$$

This little notation means we're interested in how some expression or function behaves as we substitute various numbers for x, especially as we try numbers closer and closer to 100. The thing we're interested in observing is likely to be a function. By itself that little notation isn't a math problem, or a set of instructions, or anything very useful at all. It just describes the boundaries of whatever follows it. It doesn't tell us how fast we're driving, only the speed limit. If we are using the limit process to observe a function, we put the limit information first, followed by the function we're interested in.

The dinner tip is fifteen percent of the total bill:

$$(\text{total bill})(.15) = \text{tip}$$

Expressed as a function, the same thing looks like this:

$$f(\text{total bill}) = (\text{total bill})(.15) = \text{tip}$$

This means whenever we see $f(\text{total bill})$ we know we're going to be multiplying the total bill times fifteen percent.

To describe the process of finding the correct tip for a meal that costs \$123 by using increasingly close approximations we'd write:

$$\lim_{\text{total bill} \to 123} f(\text{total bill}) = \text{real tip}$$

We would say, very carefully, "The limit of the function of the (total bill) as the total bill approaches $123 is the correct tip." The limit will be the answer to a question, perhaps one that we can't legally answer with algebra. If we want to know what x equals when something else is divided by zero, algebra prohibits us from asking the question directly. We get around this by saying, okay, we're not going to divide by zero, but we're going to divide by the smallest fraction of a number we can imagine that isn't zero. If our answers get closer and closer to some number as our estimate approaches zero we have side-stepped the laws of algebra and extracted the information. The limit process is a mathematical loophole, a slick little trick, but it is also the idea that makes all the rest of calculus possible.

As we try numbers closer and closer to 123 we'll see that the result of f(estimated total bill) gets closer and closer to $18.45. When we decide we're close enough for our purposes, we "go to the limit" or "take the limit" and declare the game over.

As you learn the limit process, you will study equally absurd examples, where the answer is obvious and simple, and the limit process is clearly a waste of time. Enjoy this time. It's your chance to become familiar with the concept while using simple examples. Later, when the answers are much less obvious, you'll be glad you spent the time. If you decide to be too clever at this point, say by merely multiplying the dinner bill by fifteen percent, you'll save lots of time now, but you'll be completely lost later.

Continuous Functions

The limit game, fun as it is, has its own limitations. On some roads you can plot a catmobile's location quite accurately using it. But if the road dead-ends, or is interrupted by an unexpected cliff, it's useless. If the restaurant automatically adds a tip to dinner bills over a certain amount you will get goofy answers. All your approximations below that number approach the correct tip, but all the numbers above it will tell you to leave zero dollars as a tip.

The limit process only works on predictable functions that don't have unexpected weirdness in them. If each change in one variable (like time) causes a specific and predictable change in the other variable (like location), the limit process is likely to work. Small changes in input (called the domain) create small changes in output (called the range). On the graph, as x moves along the x axis, the y variable will not skip up and down wildly but will move without jumps. So for the function $f(x)=x^2$ we'd write

$$\lim_{x \to a} x^2 = a^2$$

No surprise that as x gets closer to a, x^2 gets closer to a^2. In fact, one wonders why did we bother with the limit? Just to demonstrate how it works. As the unknown approaches some spot, the limit approaches the correct answer. If the function is to square something, and x is the quantity we're squaring, as x approaches 2 the limit is obviously 4 because 2^2 is 4. Functions that behave like this are called continuous functions. The graph made from a continuous function will have one nice continuous line and no gaps or jumps. In fact, you might get the idea that the definition of a continuous function is one you can graph without lifting your pencil from the paper. That may well be true, but it's a side effect. Continuous functions, and all the other functions, are indifferent to what you do with your pencil. They're continuous because the limit process works on them. It works on them because they are consistent and predictable regardless of the variables. Here "works" means that finding the limit of the expression gives you the same result as plugging in the number. The graph made from a continuous function will have one nice continuous line with no gaps. Abrupt angles may result from graphing continuous function, but they won't be differentiable functions.

Some functions behave nicely most of the time but have a spot or two where they do not. Their graphs will have a gap, for example. Since calculus depends on the limit process, it requires continuous functions. Sometimes we get around this by confining our observations to the portion of the function that *is* continuous. As long as we avoid the cliff, we can map the road. A function is continuous if our answers get closer to the correct tip as our approximations approach the actual dinner bill.

Not all functions are continuous. A chaotic function has the property that small changes in its input variable (or domain) creates unpredictably large changes in the output (or range) variable. Most of calculus and its parent field "analysis" studies continuous functions. The field of chaotic functions is relatively new and non-standard. You'll be delighted to learn that the folks who are working in this area are creating lots of brand new and fun mathematics.

$$f'(x) = \text{calculus}$$
$$\text{algebra}$$
$$f(x) = x^2 + x + 1$$
$$2x + 1 \neq 0$$

The Actual Process of Calculus

To actually use calculus, you'll assemble information from the real world and figure out a way to describe it as an algebraic expression. There are only a few forms this expression can actually take.

Then you will stare at this expression for a half hour or so until you decide which of the calculus processes apply to your particular creation. Is the whole function squared? Perhaps you'll want to use the power rule. Does the whole mess look like a fraction? You might consider the quotient rule. Often you will use several of these processes in succession, since your lovely little masterpiece may well have more than one type of function within it, each one hungering for a different procedure.

After you've plotted your strategy, you will choose one or more of these little processes to transform your algebraic expression into a derivative. The derivative is a new expression which describes the rates of change of the two variables regardless of their specific values.

Once you have done this you use the derivative (which will look like any other plain old expression) in an algebraic equation of your choice (depending on what exactly you want to learn) and then solve it using the techniques you learned in algebra.

Differential Calculus is learning these few rules and when to apply them. This is what you do to solve a problem in differential calculus:

Step One

Describe the real world information as a function or a combination of functions. In a calculus course, this will often be done for you by the authors of your Required Calculus Textbook.

wow

Step Two

Decide which rules of differentiating apply to this function or combination of functions. If the function itself is squared or raised to some other power, you may choose the power rule. If you are adding two functions together you'll use the sum rule. There are specific procedures for exponential functions and trig functions. Often the choice will be obvious.

Step Three

Use the rule itself. Most of these are quite simple, some are a little more complex. The rules tell you how to change the function, or combination of functions, into its derivative. Most of these rules do not involve calculation, but are just procedures for changing the function itself. This is the great secret of calculus: the process doesn't even feel like math, it's just a set of easy little procedures to perform on a function, things like "reduce the exponent by one and put the old exponent in front of the whole mess." In fact, that's one of the rules right there. Most of your calculus experience will be learning and practicing these little procedures.

Step Four

Create an algebra problem that uses this derivative to answer the question you're asking. For example, you might set the derivative equal to zero. Often, in calculus class you won't get to do this one.

Step Five

Use algebra to solve the problem you just created. Obviously, people who really understand algebra have a big advantage in calculus class. You won't get to complete problems as much as you'd probably like, however, since the focus of the class will be on finding derivatives.

Cats Graph Derivatives

Catmobiles have a special headlight that shoots out a narrow laser light beam in a perfectly straight line. They have an identical tail light shooting a laser beam from the rear bumper.

If the headlight hits you, you know the cats are heading straight for you. If the tail light hits you, you are too late, the cat has escaped once again.

When the vehicle is exactly horizontal, sitting on a flat plane on the cat's pancake-shaped home planet, the laser beams never hit the ground. They stay parallel to the plane.

The headlight and taillight make a straight line that is inclined exactly the same as the vehicle itself, which is the same as the slope of the road it's on. The tilt meter inside the catmobile will register exactly the same amount of inclination as the head and tail lights.

We could track the changing slope of the catmobile by building a wall in front of the road. The headlight will make a little spot of light on the wall and the spot will move up or down as the slope changes. There are a couple of especially interesting special cases.

If we're heading horizontally on the cat's home planet, the spot of light doesn't move. The slope is zero.

If we happen to be heading straight up like a rocket, the rise is huge and the run is zero. Since we can't divide by zero, we say the slope is undefined.

Places on the graph where the slope is either zero or undefined are called critical points. The graph is either horizontal or vertical at these points.

If we're driving up or down a steady incline, the spot of light doesn't move.

But if the road gets steeper for a moment, the light moves. If it flattens out, the light moves again. How high or low the light spot appears on the wall indicates how steep an incline the catmobile is on. It moves up or down when the amount of steepness changes.

If the steepness is constantly changing, the light will be moving up or down right along with it. For example, as we drive over a hill that is a perfect semi-circle shape, the light from our headlight will move downward the entire time, regardless of whether we're going up one side of the hill or down the other.

If we drive through a semi-circle shaped valley, and we've arranged for our laser to penetrate the dirt, the spot of light from our headlight will move continuously upward.

Whenever the light changes direction, from moving up to moving down, the shape of the road has just changed from a valley to a hill. When it reverses again so that it's now moving up, we're going through another valley. The places where the light spot changes direction are called "points of inflection."

If we want to know where the catmobile is at any instant, and we know where the headlight and tail lights are shining, all we need is one more coordinate. We either need to know its altitude or how far it is away from the origin. Then we can draw a triangle and use simple geometry to figure the location.

As the slope becomes very steep, you're going to need tall walls to shine your headlights onto. As the catmobile becomes nearly vertical, those walls have to be huge. In fact, if the catmobile ever gets perfectly vertical, you can't put your walls close enough together for the light to ever hit them. The headlight will just shine straight up between them and never register at all before disappearing into space.

Now suppose that we put a big sheet of paper on the wall. The laser headlight is set so that it will darken the sheet of paper wherever it hits. If the slope never changes, we'll get a black spot. If the slope does change, we'll get a vertical line.

But just looking at a vertical line doesn't tell us much, so we arrange for the paper to move from left to right at a steady speed. Now we can record whether the light was moving up or down. If the slope of the catmobile changes quickly, the line it makes will be steep.

The track the light makes probably won't look much like the shape of the road, because that's not what it follows. It graphs the changing steepness of the original road.

(To make the record more precise, we're going to assume that our wall is moving away from the origin at the same rate as the cat-mobile, so it's always exactly the same distance from the headlight. Otherwise, the angle of the lights will affect the reading a little as they get closer to the wall. Or we assume the wall is so far away it doesn't matter. At first we weren't going to mention this, on the theory that this is cats and headlights, it isn't rocket science. Then we realized it *is* rocket science, or at least the math that makes rocket science possible. So, our walls are moving away from the origin at the same rate as the catmobile. Don't send us angry mail.)

The graph our headlights make is a record of the derivative of each point of the road. In your mind's eye, gazing up at that brown line etched across the paper, you are staring at what a derivative looks like. In fact, it's the graph of the derivative function.

Let's build a new road that matches this line on our sheet of paper and drive a catmobile over it. Now our headlights will graph an entirely new shape. This new graph will chart the changing slope of the derivative itself. We call this new chart the "second derivative." If we create still another road from the second derivative and drive the catmobile over it, we chart the third derivative. Each subsequent chart is likely to be different from the one it's made from. If our original road was plotted by a polynomial function, each new graph is also likely to be simpler than its predecessor, until we are graphing a straight horizontal line.

Let's go back to our original road and the graph of its derivative. We can tell whether we were going over a hill or through a valley by the direction the graph of the derivative is heading. By comparing the graph of the headlight and the tail light we know when the slope is steep and when it's horizontal. When it's horizontal, the tail light will shine exactly as high on the wall behind us as the headlight does on the wall before us. As we crest a hill there will be an instant when that happens. This spot is called the maxima. When we hit the bottom of a valley, the head- and taillights will also be at precisely the same height. This is called a minima. If we combine the headlight and tail light charts, the two graphs will intersect at this spot.

If we all had catmobiles and big walls we could do calculus without using algebra at all.

Since we don't, we have to work a little harder.

Since any information that can be graphed as a line can be expressed as a function, we translate our information into a function.

Then we use one of the processes of calculus to convert this function into its derivative.

Then we create an equation which uses this derivative to locate whatever spot on the curve we're interested in.

Then we solve the equation using the techniques of algebra.

The Process

know

The Rules of Differentiating

When you describe the real world in terms of functions, then combine them into an expression, the expression you finally write down will fall into one of a very few categories. To find the derivative of the expression, you simply have to apply the correct procedure or "rule" to it. Here are the forms the expression can take:

1.   An expression might consist of some function plus a constant. The function might be x^2, the constant might be +35. The whole thing might look like $x^2 + 35$. To deal with the constant that is added (or subtracted) you'll use the "**Constant Rule.**"

2.   An expression might consist of a function multiplied (or divided) by a constant. If so, you'll use the "**Constant Multiple Rule.**" So if you see $35x^2$, you'll use this rule.

3.   If you've got several functions that you add together (or subtract from each other), like $x^2 + (x - 1)$ you'll use the "**Sum Rule.**"

4.   If the functions are multiplied times each other, like: $(x^4 + 34)(4x - 12)$, you'll use the "**Product Rule.**"

5.   If you divide one function by another, you'll use the "**Quotient Rule.**" You'll use the quotient rule to differentiate something that looks like this:

$$\frac{x^5 + 15x}{65 - x^2}$$

6.   If you've got a the variable x which is squared or raised to some other power, you'll use the "Power Rule." If the expression is x^{34}, use the "**Power Rule.**"

7.   Finally, if a whole function is "inside" another function you'll use the "**Chain Rule.**" This applies whenever the output of one function is used as the input for another function. For example, perhaps a trig function is operating on some other function. If you see: $\sin(x^2 + 35)$ you'd use the chain rule.

That's it. You'll still need to know how to take derivatives of several special functions like the trig functions, logarithmic and exponential functions. But you'll always use these seven rules to organize the process.

Translating the real world into functions isn't always difficult, it's something you probably did in algebra class. Just because you call something a function doesn't necessarily mean its a scary and complex beast. Doubling something is a function, and it looks like: $2x$. If you were on your own personal mental field trip when the rest of the class created functions from events, it's not too late to learn, but you'll be doing it on your own time. In most calculus classes you'll spend more time fooling with problems someone else has created. It's the efficient way for one instructor to get 20 students all working on the same thing, but it might make calculus seem distant and removed from anything you'll ever do. Think of the exercises and graphs and equations as the many scales you must play on your instrument before they invite you to Carnegie Hall. Then secretly create some little problems to test out the ideas. Take out your trombone and squawk away to your heart's content. Ultimately, it's the only way you'll learn to play. You can't break calculus by blowing too hard.

Combining functions appropriately is mostly common sense. Perhaps you're considering the growth of your cat toy collection. Several forces may be at work here. There's the increase caused by your human buying you toys, certainly. Another factor might be the toys you steal from the neighbor cat. Then there's the toys you find in the storm sewer. You may also have to consider dog losses. After you describe each of these as a function you'll need to add the functions together to predict your total toy collection. To find the derivative of this expression, you'd use the sum rule.

When you describe rates, you probably instinctively create fractions. Miles per hour is miles divided by hours, wages per hour is wages divided by hours. It seems logical to you because it's reversible. Once you have the rate, you can figure distance traveled by multiplying that rate by the amount of time traveled, which is the reverse operation of division. A rate is one kind of fraction, and it seems the natural way to express the relationship between some things. Similarly, when you're combining functions into expressions, sometimes they'll take the shape of a fraction. When they do, you'll use the quotient rule to find the derivative.

Sometimes your expression will involve functions that must be multiplied times each other. Perhaps you are manufacturing cat-mobiles and your profit on each one is a function involving variables of labor cost, materials cost and price. The number of sales might be a function that involves the price and how many days prior to Christmas it is. Your total profit would be the profit per car times the number of sales.

(profit per car)(number of cars sold) = total profit

74

Since this is a multiplication problem you'll use the product rule to determine the derivative.

Clever cat that you are, you already see the value of using calculus in this way. Increasing the price will cut sales but improve your profit per car. What price gives you the maximum total profit? To find out, you use the product rule to find the derivative. Then you'll create an equation in which the derivative equals zero. This is the mathematical way of asking the question, "what number equals x when the slope of the graph is zero?" That is, what is x when the function is at a maxima or minima? With a couple other quick calculations you can tell which it is and you'll know the ideal price for your product. Cats who use calculus like that earn huge salaries.

When you describe area, or volume, or things that accelerate, you often wind up with unknowns that are squared or cubed or raised to some other power. Sometimes an entire function must be squared or raised to a higher power like this:

$$(x^2 + 1)^4$$

To find the derivative of an expression in this form you'll use the chain rule.

Constant Rule

As you begin calculus, you'll be trying to determine what y equals for some x. You want to know the specific vertical height of the graphed function when it's directly above some spot on the x axis. You can do this because y is a function of x, or, as your math instructor will say:

$$y = f(x)$$

This is so obvious to the folks who write calculus books they will neglect to mention it, or will deal with it quickly at the beginning of the book while you're still too confused to remember it. So I'll repeat it here: Our goal is usually to deduce what *y* represents for some specific *x*.

If we start with 10 cans of cat food and buy one can every day, we could chart our cat food progress with a little chart like this:

day	number of cans
0	10
1	11
2	12
3	13
x	$x+10$

We might express this as the function:

$$f(x) = x + 10$$

Being infinitely bored, we want to use calculus to predict our future food supply. That means we want to know the slope this expression would graph on any day. We want to find its derivative.

If you pick days at random and plot a line between them you'll quickly see that you draw a nice straight line which has a slope of one. That is:

$$f'(x) = 1$$

What happened to the 10 cans you started with? Turns out they had no effect on the slope whatsoever. If you started with 35 cans, or a thousand cans, they still would have had no effect on the slope. They give you the starting point, but don't affect the steepness of the line. Any fixed number is called a constant. The message here is that, when seeking derivatives, we can disregard a constant that is added or subtracted to a function. In fact, we can simply cross it out. It represents how high above (or below) the x axis you started graphing your function, but not the slope of the line you draw. We bought a can each day, and consumed one each day. Our total store of cans remains unchanged.

On the other hand, what if our function turns out to be a constant all by itself? That is, if we can simplify the function down to a fixed number, like 10 or 35? If we can, the function is graphing a straight horizontal line. The derivative of a straight horizontal line is zero. So, if you're asked to find the derivative of 653, the answer is zero.

Constant Multiple Rule

If a function contains a constant multiplied times the rest of the function, you don't need to fool with the constant. You simply transplant it whole into the derivative. That's the Constant Multiple Rule.

The derivative of the function $35f(x)$ will be 35 times the derivative $f'(x)$.

Regular old numbers are constants. Things like square roots of numbers are also constants, because they're just another way of describing some number. The famous number pi is a constant for the same reason. This little rule lets you deal with some formidable looking fellows.

Power Rule

Often a function includes x squared or cubed or raised to some other power. Other branches of math cringe in terror at these because they can involve large numbers and lots of calculations. Calculus treats them with the casual disdain of an only cat.

To transform a function that's raised to a power into its derivative, you multiply the function by the number of the power and reduce the power by one.

If the function is x^{45}, its derivative is $45x^{44}$.
If the function is

$$x^{103}$$

its derivative is

$$103x^{102}$$

This is different than

This, of course, is way too easy. Because you have taken Algebra I, Algebra II, Geometry and Trigonometry, you don't really believe something as simple and effective as this without performing all the steps that prove it to be true. Rest assured, your calculus instructor understands your need for thoroughness and will give you ample opportunity to satisfy yourself. We would not want to spoil that fun by doing it here.

There's an interesting special case of using the power rule you'll come to love. You can think of x as x^1. If you do, and you want to differentiate it, use the power rule. We multiply the function by the number of the power, which is one, and reduce the power by one. That gives us $1x^0$. Since anything raised to the zero power is one, the derivative of x is 1.

If the function is x raised to a power in the denominator, you can rewrite the function so it's no longer a fraction, but rather x taken to that negative power. This algebraic trick transforms:

$$\frac{1}{x^2}$$

into x^{-2}, which is much easier to keep in its cage. If the functions is x^2, its derivative is $-2x^{-3}$.

Sum Rule

If an expression contains functions added together, you find the derivative of each one and add them together. That's the sum rule.

We start with two functions added together:

$$(x^4 + 12) + (x^3 - 5x^2)$$

We take the derivative of each. In this case, each of the two functions is also an addition or subtraction problem, so to find their derivatives, we use the sum rule on them as well. We find the derivatives of each part of the addition problem by using the power rule and the constant added rule:

derivative of $(x^4 + 12)$ is $4x^3$
derivative of $(x^3 - 5x^2)$ is $3x^2 - (5)(2x)$ *(mae)*

We add up the two derivatives. By using the sum rule, we see the derivative of the original expression is

$$4x^3 + 3x^2 - 10x$$

Then combine

The sum rule applies no matter how many functions you're adding together. Despite its name, it also works if you're subtracting functions from each other, or adding some and subtracting others. Subtracting is the same as adding negative numbers.

To find the derivative of this:

$$f(x) = x^{15} - 23x + 16$$

we use the power rule on that first function (which is x^{15}) and come up with:

$$15x^{14}.$$

In the second function (which is -23x), we get rid of the x with the power rule and transplant the constant to the derivative using the constant multiple rule to get -23.

In the third function, we simply eliminate it using the constant rule to come up with +0.

Using the sum rule, we declare the derivative of the whole expression to be

$$15x^{14} - 23 + 0$$

As an abbreviation for the entire preceding sentence we write:

$$f'(x) = 15x^{14} - 23$$

You can now take the derivative of any polynomial using combinations of these rules.

Until now, we've been coasting. The rules are so simple and easy to grasp that you've probably got them memorized. There are only three more rules, but each of them requires a bit more effort.

Recall our favorite expression that involves multiplying two functions:

(profit per car)(number of cars sold)

The product rule applies to expressions like this. It involves three steps.

1) First we find the derivative of each function separately;

2) Then we multiply each function by the derivative of the other;

3) Then we add the results together.

Think of it this way:

the expression is: (first)(second)

the derivative is

(first)(derivative of the second) + (second)(derivative of the first)

A practical tip: When you start seeing fearsome expressions it's easy to lose track of what you're doing. To avoid this, it's wise to develop a habit of copying the first factor, then find the derivative of the second, then copy the second factor and find the derivative of the first. If you remember "copy first, take derivative of the second, plus copy second, take derivative of the first" you'll be less likely to lose your place.

Step One

Copy (profit per car)

Find the derivative of (number of cars sold)

write a plus sign

Step Two

Copy (number of cars sold)

Find the derivative of (profit per car)

Step Three

Multiply (profit per car) times the derivative of (number of cars sold)

Multiply (number of cars sold) times the derivative of (profit per car)

Step Four

Add these two results together.

It would be really handy if you got this firmly into your head before your instructor starts using actual beauty-impaired functions constructed of real numbers, x's, fractions and other evilness. The concepts seem easy the way a lovely martial arts kick does when you practice in front of your mirror. It's amazing how much less graceful we feel when cornered by a knife-wielding expression dressed in exponents and square roots. When you're actually backed into a corner and forced to defend yourself against actual calculus problems, you are less likely to make a mistake if you remember this "copy first, take derivative of the second, plus copy second, take derivative of the first" mantra. Back to the mirror for one more example:

You are describing some lovely woodland dance that x will perform. This dance is a combination of two other dances, the Bambi Trot and the Hokey Pokey. Each of these dances is a function. If we were doing one dance then the other we might decide to add the two functions. But we're not. We're really doing both dances at the same time. We may have to look around us quizzically or stare into headlights (for the Bambi Trot) at the same time we're putting our left foot out or shaking it all about (for the Hokey Pokey). This is not a sequential process like adding, since we don't do one then the next. It's not a rate which we could express as a quotient. We're not raising anything to a power. Perhaps we decide the only appropriate way to combine them is to multiply them times each other.

$(x$ does Bambi Trot$)(x$ does Hokey Poky$)$

To find the derivative of this expression we'll use the product rule.

We use the power rule, or the constant rule or whatever is appropriate to find the derivative of each function, which we'll designate in this example by adding an apostrophe.

Derivative of (Bambi Trot) looks like (Bambi Trot)'
Derivative of (Hokey Pokey) looks like (Hokey Pokey)'

We multiply (Bambi Trot) times (Hokey Pokey)'
We multiply (Hokey Pokey) times (Bambi Trot)'
Then we add the two products. Their sum is the derivative.
The only way we could possibly make this confusing would be to use numbers, exponents, and unknowns, much as they do in Required Calculus Books.

For example:

$$f(x) = x^{14}(35x + 12)$$

Step One

Find the derivative of x^{14} which is $14x^{13}$
Find the derivative of $35x + 12$ which is 35

83

Multiply each function by the other's derivative:

$$x^{14}(35)$$
$$\text{and}$$
$$(35x+12)(14x^{13})$$

Step Three

Add them together:

$$14x^{13}(35x+12) +35x^{14}$$

You can simplify this, or complete the math, or use it as it comes. This is the derivative of the original expression.

The product rule works on expressions that contain more than two functions. Multiply the derivative of each function times every other function and add the results.

If the expression is:

$$(rat)(cat)(bat)$$

the derivative will be

$$((rat)'(cat)(bat)) + ((rat)(cat)'(bat)) + ((rat)(cat)(bat)')$$

A Brief Aside—Why Poets Don't Do Calculus

Some people like things to be precisely accurate. If you say, "I'll be home at 6," these folks do not think 7:30 is close enough. If they are either your mother or your wife, they may not even think 6:30 is close enough.

Most mathematicians think of themselves as being this kind of person; they love the precision of math. But calculus has some imprecision built right into it and they feel guilty about it. Rather than just say that some of our answers will be a tiny bit off, but that this infinitesimal error doesn't matter for our purposes, they want to convince you their extremely close approximation is so close to the truth that it's not a lie. They want to prove to you that .99999... is exactly the same thing as 1, for example. They won't feel so bad about living this lie if they can convince you to join them in it. Of course, those of us with English Lit backgrounds don't understand their guilt. We say "my love is a rose" with little provocation, regardless of how rose-like our love is, and we think "to be or not to be" is actually a question. We won't make you spend weeks doing problems to convince you that our love actually is a rose, and we don't really feel bad about it. But then, we never pretended that poetry was based on consistency and precision. In math class, you must come to agree that .9999... actually is identical to 1, and until you wear that saddle peacefully, they won't let you roam the pastures alone.

The product rule gives us an easy way to see where this imprecision slips in, and why we can live with it. If you retrace Leibnitz's process, you can describe the derivative of xy (a product) like this:

$$d(xy) = (x + dx)(y+dy)-xy$$

That is, (*x* plus the tiny change in *x*) times (*y* plus the tiny change in *y*), then subtract out the original function to leave only the change in the overall product.

After you do the algebra, you see that the derivative is:

$$(x + dx)(y + dy) - xy \qquad xdy + ydx + dxdy$$

That is, the first factor times the derivative of the second, plus the second factor times the derivative of the first, plus this other thing, the two derivatives times each other.

What Leibniz decided is that you could simply discard that last *dxdy*, on the theory that since we were talking about two infinitely small quantities, multiplying them together resulted in something that was so small as to be insignificant.

And, of course, he was right. We're throwing away the tiniest smidgen of change, and it can't possibly affect anything.

We do suggest, however, that if you're ever using calculus to create a model of the universe, or of sub-microscopic particles, or a theory of reality, or anything else where you're extending data into huge numbers, and assuming that your math models reality perfectly, don't forget that you're going to be off by this little smidgen. If you do, your time travel machine is going to think that .9999... equals 1, which could be the difference between Pittsburgh and the Pleistocene.

Quotient Rule

We use the quotient rule to find the derivative of expressions that look like fractions.

$$f(x) = \frac{top}{bottom}$$

First we find the derivative of the numerator and denominator by themselves.

The derivative of (top) will be (top)'
The derivative of (bottom) will be (bottom)'

Now we have four different characters to play with, top, top', bottom, and bottom'. To find the derivative of the original expression you simply plug the appropriate one into the right spot in this charming formula:

$$\frac{(bottom)(top)' - (top)(bottom)'}{bottom^2}$$

That's the quotient rule. There's nothing tough about it except understanding it and remembering it. Your calculus instructor wants to be the person who helps you understand it. We'd like to help you remember it.

As any cat owner knows, two very common functions are the "poop" function (abbreviated P) and the "oops" function (abbreviated O). When you are trying to remember what to do with the quotient rule, we humbly suggest you think of it in this form:

$$\frac{O}{P}$$

We call this the "Oops over Poop" form. To find the derivative, you simply do this:

$$\frac{PO' - OP'}{P^2}$$

Luckily, this spells out "poop-over-poop-squared," which is easy to remember. All you have to do is remember which ones are derivatives, and that you subtract in the middle of your upper poop. This may be our proudest contribution to the world of mathematics.

Another way to think of this is to consider the function:

$$mess = \frac{Oops}{Poop}$$

We can use basic algebra to transform this into:

$$Oops = (Poop)(mess)$$

To translate that into language your math instructor might accept, we'll use z to represent the mess:

$$x = yz$$

Then we can use the product rule to discover that:

$$dx = ydz + zdy$$

If we solve for dz:

$$ydz = dx - 2dy$$

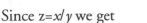

Since $z = x/y$ we get

$$ydz - dx = \frac{-xdy}{y}$$

$$dz = \frac{dx}{y} - \frac{xdy}{y^2} = \frac{ydx - xdy}{y^2}$$

Your math instructor will probably prefer this way of expressing the quotient rule.

Chain Rule

To find the derivative of an entire function that is squared or raised to some other power (or which is acted upon by some other function, like a trig function), you'll use the chain rule. The "inside function" will usually be enclosed within parentheses. The outside function will be the exponent or trig function that you apply to the inside function. In this:

$$y = (x^2 + 1)^3$$

(x^2+1) is the inside function. Cubing it is the outside function.

The chain rule involves three steps. Always work from the "outside" and move inward.

Step One:

First use the power rule to find the derivative of the whole expression.

To find the derivative of $(x^2 + 1)^3$
first we bring down the $\qquad$ 3
then copy the inside, $\qquad 3(x^2 + 1)$
then lower the exponent by one $\qquad 3(x^2+1)^2$

Step Two:

Then find the derivative of the inside function. In this case, we use the power rule again. We bring down the exponent, copy the function, lower the exponent by one. We also use the constant added rule to eliminate the +1:

The derivative of $x^2 + 1$ is $2x$

Step Three:

Then multiply these two derivatives times each other:

$$3(x^2+1)^2 \text{times} 2x$$

After you neaten it up with algebra, you learn that the derivative is $6x(x2+1)^2$

The inside function might be a wild and hairy looking beast, intimidating and inconvenient to rewrite. The natural solution for the mathematically inclined is to abbreviate it with a letter. But most of the good letters have already been taken. What to do?

The Use of Letters in Math

The letters near the beginning of the alphabet *(a, b,* and *c)* staked their claim years ago to representing constants, like the lengths of the sides of a specific triangle.

The letters near the end of the alphabet like *x, y* and z have claimed the right to be unknowns.

The letters near f typically represent functions. If there is only one function in an expression it's likely to be abbreviated *f*. If there are two, they will probably be *f* and *g*. A third function will probably be *h*. The letters i, j, k, l, m, and n are referred to as the "in crowd" because they're the letters of the alphabet between i and n. Mathematicians usually think of them as representing integers. Some of them are used in even more specific ways. For example: n stands for any number when you're describing some process like x^{n-1}

m refers to the slope of a straight line in algebra. Small m replaces n if you've got a process that requires two numbers and you already used n.

The letter d represents the tiny change in the *x* or *y* coordinates that determine the slope of a graph we're interested in.

The letter e is associated with natural logarithms and exponential functions. It stands for a very specific irrational number, 2.718....

Time is involved so often in equations that we reserve the letter t for time.

Width is w.

r stands for radius or radian.

s indicates position.

O stands for "oops" in this book. It looks too much like a zero to use otherwise.

P stands for a point.

V is used for volume, small v for velocity.

K is the abbreviation for Kelvin, a scale of temperature. It also represents a fixed value in differential equations.

Small L looks a lot like a one

Q also looks like a zero; the lower case q looks like a g.

That leaves us with the letter u for no logical reason except it's all that's left. The inside function is typically abbreviated with the letter u.

The chain rule works any time you have an inner function being operated upon by another function. If the outer function is to take the square root of the inner function, we'll rewrite the expression as a power, then use the chain rule. It's easy to get lost in this process of taking derivatives, then copying something from the problem, then taking another derivative. By abbreviating the inside function with the letter *u*, we reduce our chances of losing our place or getting confused until we need to produce the actual beast. At that point, we replace *u* with the real expression. The whole thing looks a lot simpler. In our example:

$$y = (x^2 + 1)^3$$

the whole thing looks a lot less frightening if we abbreviate $x^2 +1$ (the inside function) with the letter u:

$$y=u^3$$

When we've gone through all our derivative games we get:

$$y'=3u^2 \text{ times } u'$$

We haven't changed the meaning of anything, just simplified the way it looks.

Trig Functions

Sometimes cats do things that seem silly to the untrained observer. For example, just as many humans pace the floor while deep in thought, deep-thinking cats often climb things. One can't help but wonder how many feline Einsteins and Descartes have had their concentration shattered by the surprise and inconvenience of a Christmas tree toppling beneath them. How many important scientific and philosophical insights have been lost this way?

One such "felosopher", a brilliant cat named Binkie, climbed a tall clock tower while deep in complex abstract thought. Without noticing what he was doing, he somehow managed to get out onto the clock face itself, high above the city, and climb to the tip of the two-story-long second hand. Only after he had completed his deep thinking did he look around, realize his predicament and begin to yowl. He dug his claws in tightly as he rode around and around the face of the clock, unable to back down. There was only one hope of rescue. He would have to summon one of the cats' top secret black helicopters.

Unlike the catmobiles, the black helicopters are hard to hide, but because they are perfectly silent and black, they're almost impossible to detect at night. Unfortunately, Binkie could not wait until dark. Although he was brilliant, he had let his clinging skills deterio-rate and his claws were already beginning to tire. Luckily, cats have developed a foolproof technique for such situations. Binkie would have to calculate what his precise position would be at the instant the helicopter arrived. The vehicle would hover for a single second at that spot, he'd leap aboard, then it would speed away. By the time an obser-vant human saw it, reacted and pointed, it would be gone. Any humans who looked where he pointed would see nothing. The human who pointed would be branded a lunatic and Binkie would be safe. It was a perfect and very cat-like plan.

Binkie knew it would take exactly 1,643 seconds for the heli-copter to arrive. But where would the second hand be at that moment? More importantly, how fast would it be going up or down? In order to make the leap, the vertical speed of helicopter and second hand needed to be precisely identical. Binkie was a proud cat with a deep love of calculus. He determined to convert the motion of the tip of the second hand into a function and perform one elegant calculus calculation to solve his dilemma.

There are several equations that will graph a perfect circle. Unfortunately, the most common one doesn't result in a single unique y for each x. It's not a "one valued function." For most points on the horizontal axis it graphs two points on the vertical axis. At the origin, for example, (the center of the clock face), a vertical line would inter-sect with both 12 o'clock and 6 o'clock. The one o'clock mark is directly above the five o'clock mark. What Binkie needed was an equa-tion that would graph the equivalent of his nightmarish trip around the clock but that would not graph a circle. He needed some function that graphed an analogy of the tip of a second hand moving around a circle.

It occurred to him that the angle of the second hand was changing in a precise relationship to his location. He imagined a horizontal line from the center of the clock face to the three o'clock mark. The angle of the second hand to that line changed each second and never repeated itself in a single revolution. Perhaps Binkie could describe a circle in terms of that changing angle. He felt certain he was on the right track.

But the concept of "angle" has no meaning in the Cartesian sub-world of algebra or calculus. If we're thinking about points being so far up or down from the x axis and so far left or right of the y axis, telling us that one line intersects another at 10 degrees tells us about as much as telling us it tastes like sauerkraut. It has to be translated into the system. The Babylonians divided a circle into 360 "degrees" and we've been using their idea ever since, but they could have used 400 or a thousand divisions just as easily. If they had used other numbers, then what we call ninety degrees might be called "100 degrees" or "250 degrees." Measuring angles in degrees has become so common we need to deal with it, but it's an arbitrary unit. To be useful in calculus we have to describe angles in terms of some process or relationship. We need to use the angle itself in a function.

Child's play for a cat. Binkie mentally drew a line straight down from his position to his imaginary horizontal line. Now he had a triangle consisting of the horizontal line, the vertical line and the second hand itself. Each time the second hand moved it created a new triangle with slightly different dimensions.

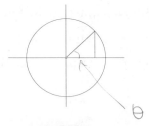

The sides of each triangle have an exact and unique relationship to each other for each angle of the second hand. That relationship is a different way to express the angle. Any angle can be expressed as a ratio between the lengths of the two sides. It looks like a fraction, it acts like a division problem, and as luck would have it, it's a continuous function. If Binkie plotted the change in those relationships using time as his *x* axis, it wouldn't graph a circle at all. For example if he plotted the relationship between his altitude above the horizontal line and the length of the second hand it would look like this:

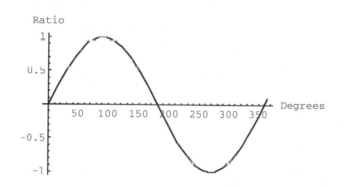

This relationship is called "sine" (rhymes with "mine"). To demonstrate their love of abbreviation, mathematicians abbreviate "sine" as "sin" when actually used in an equation, but they don't pronounce it the way it looks. It still rhymes with mine. There are six possible relationships between the three lines of a right triangle, and each one has been given a Greek name. The angle itself (between the second hand and the imaginary horizontal line to three o'clock) has been given the Greek name theta, which is a letter in the Greek alphabet that looks like this:

$$\theta$$

Whenever you see a theta, someone is referring to the actual angle, which could be sixty degrees (or whatever) when it's translated into English.

If you know one angle of a right triangle and the length of one side, you can figure out the length of any of the other sides, because people have spent hours and hours figuring out these relationships precisely and recording them in tables and computers and calculators. Solving problems using these relationships is what you do in Trigonometry class. Trigonometry is especially handy to determine the size or distance of something you can't measure directly. You create a triangle that includes the side you can't measure plus angles and sides that you can measure, then use the tables to help you figure what the unmeasurable side must be.

Depending on what class you're taking, and when you're taking it, the names of the three sides have varied. In Geometry class you learned to think of the sides of right triangles as a, b, and c. (Perhaps you remember $a^2 + b^2 = c^2$?) In the 1960's, taking a trig class, you called them OR, RP, and OP. I like the names they used in the 1940's myself, and they can't possibly hurt you, although you'll have to learn their new names sooner or later. For a moment, you're back in 1940:

They called the angle we're interested in (the theta angle) the vertex. That's the angle the second hand of Binkie's clock makes with his imaginary line from the center to 3 o'clock. From the vertex, a horizontal line extends which is called the "initial line" or "shadow". At the end of the shadow we find the perpendicular, which is a vertical line. Rising from the vertex (the angle we're interested in) like a ladder leaning against the perpendicular line is the "distance" or "terminal line." That's Binkie's second hand itself. I like these because I can picture them. The vertical line is the "perpendicular" which is easy because it's perpendicular to the horizontal line. The hypotenuse or "terminal line" is the ladder leaning up against the perpendicular. It's the second hand of the clock. The shadow is on the ground beneath the ladder, or Binkie's imaginary horizontal line. Of course, triangles won't always be aligned just like this, but it works to illustrate these functions.

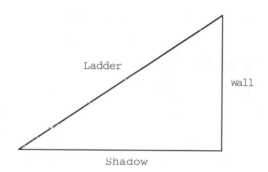

Since these guys think it's just fine to change the names of these sides, there's no reason we can't as well. For the purposes of demonstrating how easy this is, let's just use the height of the wall, the length of the ladder, and the length of the shadow on the ground as our measurements.

The sine of the vertex (or theta, or angle) is a fraction made up of the wall's height over the ladder's length;

Cosine of the vertex is a fraction made of shadow over ladder.

Tangent of the vertex is the wall over the shadow.

$$\text{Sin} = \frac{\text{Wall}}{\text{Ladder}}$$

$$\text{Cos} = \frac{\text{Shadow}}{\text{Ladder}}$$

$$\text{TAN} = \frac{\text{WALL}}{\text{Shadow}}$$

TAN =
S/W

Secant = L/W   *Cosecant = L/W*

The other three possibilities are the reciprocals of these. That is:

Cosecant of the vertex is ladder over the wall
Secant is the ladder over the shadow
Cotangent is the shadow over the wall.

L ⟋ *W*
S

If the vertex, or as we now say, the theta angle, is 20 degrees, it doesn't matter how big the triangle is. The ratios between all the sides don't change. A 10 mile triangle will have the same ratios between sides as a four inch triangle will have.

As the twenty-first century begins, the line we called the ladder is known as the "hypotenuse," just as it was in geometry class, the shadow is called the "adjacent" side, and the wall is called the "opposite" side. Many folks remember what the trig functions represent by memorizing the name of the trig cat, SOHCAHTOA.

This is useful to them because each triplet of letters in his name stand for the relationships the Big Three trig functions represent:

SOH: Sine = Opp/Hyp

CAH: Cosine = Adj/Hyp

TOA: Tangent = Opp/Adj

$$\sin = \frac{opposite}{hypoteneuse} = \frac{oh}{heck} -$$

$$\cos = \frac{adjacent}{hypoteneuse} = \frac{another}{hour}$$

$$\tan = \frac{opposite}{adjacent} = \frac{of}{algebra}$$

In calculus, the six "trig functions" (sine and the rest) are used to describe the relationship between the vertex angle and these ratios as functions. This is more useful than it sounds.

Many things in the real world are easiest to translate into math if we use trig functions. Things that rotate or spin generally lend themselves to a trig function. So do relationships that involve measuring an angle, or situations where we're interested in the direction something's heading. Navigation problems involving wind speed and direction, air speed and other variables are most neatly translated to math by using trig functions. We can also work backward from the information at hand to determine an angle. The first practical application of calculus employed this concept to figure the angle to raise a cannon so it would hit its target.

Trig functions express the changing angle of a second hand as it moves around a clock. But, of course, we need to be consistent in measuring. We don't want to measure some things in inches and some things in centimeters. Since trig functions measure things in terms of relationships rather than specific inches or feet, we track the tip of the second hand as a relationship as well. We say that one complete revolution equals 2π. Any portion of a revolution is that same portion of 2π. This simple idea is given the fancy name "radian measure" of an angle.

All the many calculations of the different triangles and ratios have been programmed into your calculator.

If you instruct your calculator to find $\sin(2.5 \bullet 2\pi)$ it will provide the ratio of the wall height over the ladder's length after the second hand has traveled around the circle two and a half times and do the calculation nearly as quickly as a cat can. If you say $\tan(35 \bullet 2\pi)$ it will calculate the ratio of the wall over the shadow after thirty five complete revolutions. You'll notice that this is exactly the same as the angle after three complete revolutions, or any other number of complete revolutions.

Trig functions look marvelously official, but it's easy to find their derivatives. The derivative of each of the trig functions is one of the other trig functions, sometimes with minor modifications. All you

99

have to do is memorize them, or come up with some clever story to remind you which is which, and you've saved yourself a lot of time.

the derivative of sin(x) is cos(x)

the derivative of cos(x) is -sin(x)

the derivative of tan(x) is sec^2(x)

the derivative of cot(x) is -csc^2(x)

The only tricky little detail to remember is that the angle x is being measured in radians, not degrees.

Kinds of Functions

Functions are divided into two big groups, algebraic and transcendental.

The algebraic functions are polynomial functions (which include sums, differences, and variables raised to a constant power) and rational functions (quotients of polynomial functions. That is, one polynomial function divided by another to look like a fraction.)

The transcendental functions are the trig functions and exponential and log functions.

As you work your way through your Required Calculus Textbook, they will try to give examples of how each rule or technique works on each of these varieties of functions. However, they may forget to tell you that's what they're doing. If you don't happen to notice the pattern on your own, you risk becoming overwhelmed by all these different ways to use the chain rule, the power rule, etc. It may help you to remember that you're learning a very few techniques, and how to apply them to these five varieties of functions. Obviously, it doesn't take a smart cat to create problems that look intimidating if you let him use trig functions. There are only a few kinds of problems, and five kinds of functions. This is a lot like getting dressed in the morning. If you have five pairs of pants, and seven shirts, you can combine them enough different ways it may seem you have an endless wardrobe. That doesn't mean it's hard to put on a shirt.

How to Use the Derivative Rules

When you are first asked to find the derivative of some algebraic monster, your first job will be to recognize exactly which kind of monster it is. As you learned the rules in your Required Calculus Textbook, you practiced catching one kind of mouse at a time. When you came to class you knew whether you were working on sums or products or quotients, and you looked forward to the experience. You knocked out those derivatives like a pro and got perfect scores on all the quizzes.

At some point, they will begin releasing unmarked mice. Shall you use the product rule? The quotient rule? Or what?

To make this choice, you need to decide what kind of an expression is running around your living room. The expression may contain exponents, multiplication, addition and trig functions. But when you consider the expression as a whole, is it at heart an addition problem, perhaps composed of several expressions (with exponents) added together, or is the whole expression raised to a power, or is some outer function (perhaps a trig function) operating on an inner function?

One handy trick is to remember the Order of Operations from algebra. If you were plugging a number into the expression, you'd work your way from inside out, if possible. You'd follow the "Please Excuse My Dear Aunt Sally" guidelines and deal with things in parentheses first, then exponents, then multiply, divide, add or subtract. The key to the nature of the expression is "what would the last operation be?" If the last step you would perform is addition, then you'll use the sum rule. If the last operation is division, you'll probably use the quotient rule.

In the army, generals make the decisions and foot soldiers do the dirty work. In calculus, you get to be both. You first put on your general's hat and decide what techniques are appropriate to use. Then you put on your foot-soldier's hat and do the work.

Difference Quotient

Imagine you are driving down the highway in your catmobile. Every mile there's a sign that indicates how far you've come: mile one, mile two, etc. Being a cat, this is fascinating to you.

If you look at your watch every time you pass a mile post, you'll gather all the information you need to create a "difference quotient," which is an expression that combines all this information into a form you can use to deduce information about your speed. This is the central idea of differential calculus.

You notice that at nine o'clock you pass milepost six, and at 10 o'clock you pass milepost eight.

Your two variables are time and location, and they're obviously related. You might be inclined to name them t and x, but if that urge does not overtake you, you can still understand the principle.

To describe the change in time, you'd subtract 9 o'clock from 10 o'clock. If you completed the math, you'd discover that one hour had passed, but we won't make you perform that math. We'll describe the time as $10 - 9$.

To describe the change in distance you'd subtract mile 6 from mile 8. Again, you could finish the math, or you can describe the change in distance as 8 − 6.

If you want to know your average rate of speed during this time, you'd create a fraction (also called a "quotient") with distance over time. We always create fractions to describe rates, but in conversation we don't say "miles over hours." We say "miles per hour." It's the same thing.

If you don't rush to perform the subtractions you come up with a fraction that looks like:

$$\frac{8 \text{ miles} - 6 \text{ miles}}{10{:}00 - 9{:}00}$$

That little expression is called a "difference quotient." The numerator is a subtraction problem(or "difference problem"), and so is the denominator. If you complete the math, you'll learn your average speed during the time span you're investigating.

In this example, you could describe your speed at any time between 9 and 10 if your speed is constant and translates into a linear function.

But, of course, your speed may not have been constant. You slowed when your radar detector started beeping at you, you sped up when you saw the sign advertising the Roadside Mouse Museum ahead. If you've taken more measurements, you may have come up with more complex functions to describe your progress.

It doesn't matter. Your difference quotient, at its heart, will still say "for every change in time, the distance changes a certain specific amount." When you ask "What fraction do we come closer and closer to as we inject smaller and smaller changes?" you are asking "What is the derivative?" In fact, that's the definition of a derivative.

In your Required Calculus Textbook, this simple concept will look much more formidable. In our example, we'll call 9 o'clock x_1 and 10 o'clock x_2. A cat would express it like this:

$$\lim_{x_2 \to x_1} \frac{f(x_2) - f(x_1)}{x_2 - x_1}$$

means dist. a function of time

If you look at it calmly and carefully, it's the same thing. The numerator describes the change in position for every change in time. It asks: "as the change in x approaches zero, what fraction do we approach?" As we look at mileposts that are closer and closer together, what fraction describes the relationship between the distance traveled and the time it took to travel it? That is, what was our speed at exactly 9 o'clock?

Once you've expressed your functions in this way, you can find the derivative with algebra and limits. If you didn't know the "rules" this is how you'd do it. In fact, it's the way they came up with the rules, and the way they prove them. When you're out in the real world, you won't usually want to spend the time to wrestle with the algebra of problems that can be easily solved by using the rules. Therefore, you may want to simply skip this entire process, but your calculus instructor has been well trained by the cats. She won't let you.

Her logic is good. She feels that, unless you go through the algebra process you'll forget that your derivative is a rate. If you get a derivative of 12, for example, the fact that it represents "12 somethings per something else" is not at all obvious. She'll believe that you won't truly believe in calculus until you've had the chance to prove each rule for yourself. She'll think you won't appreciate the rules until you've seen the hours of calculating they replace. Because she loves you, she probably won't teach you the rules at all until you've gone through this process many times. She believes that the best way to help you understand derivatives and where they come from is to encourage you to go through the difference quotient procedure until it's second nature to you. This reminds some students of the philosophy made popular by the Spanish Inquisition: "the best way to sway people to our way of thinking is to twist metal screws into their thumbs until they see the light."

The down side of the difference quotient is that you have to be good at algebra to survive it. If you mess up the algebra, you'll miss the mouse museum altogether and decide that calculus is too hard for you.

On the other hand, the down side of simply learning the rules is that you can find derivatives easily, without necessarily understanding what they represent. That's just as bad. When you run into more interesting problems, or real life situations, you'll be at a disadvantage. Just because you can find a derivative doesn't mean you know how to use it to find the cat food.

Plus, of course, this is the way it's always been taught. Do not be surprised if you show up at your first calculus course and discover that you're going to spend lots of time in the beginning wrestling with the algebra of difference quotients. This is how you will finally master algebra. Take heart from knowing that real-world calculus isn't as difficult as the algebra you'll be doing.

The difference quotient may have a slightly more generic look in your Required Calculus Textbook:

$$\lim_{\Delta x \to 0} \frac{f(x + \Delta x) - f(x)}{(x + \Delta x) - (x)}$$

We plug the specifics of our problem into this expression to find the relationship of y to x for any point we can graph.

Radian Measure

A cat named Felix stands in the center of a circular ice rink holding the end of a rope. A cat named Mambo stands on the edge of the rink holding the other end. The length of the rope between them is the radius of the circle. Mambo could skate around the circumference of that rink all day long, if Felix was willing to stand in the center, pivoting.

But cats don't have that kind of patience. Instead, they cut the rope to exactly the length of the radius and lay it around the outer edge of the circular rink. Now instead of calling it a "radius" they call it one "radian." One radian is the same as one radius.

A third cat named Sylvester passes by and cannot resist joining the scandalous good time Mambo and Felix are obviously enjoying. A new game develops.

Now, Felix stands in the center of the ice rink holding two ropes, each one radius long. Mambo and Sylvester each grab the other end of a rope and stand on the circumference of the rink. Sylvester stands still while Mambo skates around the circumference for a moment. Then he stops.

This may be too much fast-paced action for a math book, so let's review where we stand: Felix is in the center of the circle holding two ropes. The ropes form some angle with each other, depending on how far Mambo skated before he stopped.

Our task is to express the angle the two ropes make without resorting to using "degrees."

What we need to know is how far apart Mambo and Sylvester are on the circumference. Merely measuring a straight line between them won't do us any good because they skate in a circular path. Besides, maybe they're 10 feet apart because Mambo didn't skate very far, or maybe he skated all the way around and is now within 10 feet of Sylvester once again. We really need to know how far he actually skated around the rink. We lay a piece of rope along the circumference of the ice rink starting with Sylvester in the same direction as Mambo skated. When we reach Mambo we measure it. We compare the length of that rope to the length of the radius rope.

When we want to compare two things or express their relationship, our instincts tell us to divide one by the other. We divide the arc between Sylvester and Mambo by the radius. This fraction is a way of describing the angle our two original ropes made in the center of the circle.

This is called radian measure. If the arc is fifty feet, and the radius is 10 feet, we divide fifty by 10 and say the angle was five radians. If the arc was five hundred feet and the radius was one hundred feet, the angle would still be five radians. The relationship is the same, the angle is the same.

Of course, mathematicians have been studying circles for a long time and once they know a radius, they can figure lots of other things out. To them it is obvious there are 2π radians in a circle, that one degree is equal to about 0.017453 radians, and that one radian is equal to about 57.296 degrees.

The same bag of tricks tells them that it's easy to convert from degrees to radians or vice versa.

To change from degrees to radians, multiply by π over 180, or approximately 0.017453.

To change from radians to degrees, multiply by 180 over pi, or 57.296

Your math instructor probably won't ever do this, and neither will your Required Calculus Textbook. Instead, they'll confine themselves to problems that use 30 degrees, 45 degrees, 60 degrees, 90 degrees, 180 degrees and 360 degrees, because they've memorized how to convert those common angles into radian measure. Should you feel the urge to memorize something this afternoon, we'll give you that opportunity as well:

degrees	radian measure	what part of a circle is that?	on a clock
360	2π radians	full circle	noon to midnight
180	π radians	half circle	noon to six
90	$\pi/2$ radians	quarter circle	noon to 3
60	$\pi/3$	1/6	noon to 2
45	$\pi/4$	1/8	noon to 1:30
30	$\pi/6$	1/12	noon to 1

This may seem like a lot of work just to avoid some Babylonian tradition, and, of course, that's not really why radian measure has become so beloved. The strategic value of using radian measure is that when we do, the derivative of sine is cosine. If we use degrees, on the other hand, the derivative of sine is proportional to cosine, but the constant isn't 1. The derivative of $\sin(x)$ would equal $(\pi/180)\cos(x)$ or $0.01745329\cos(x)$, which is not the easiest number to use in a calcu-

108

lation. When mathematicians got tired of it, they invented radian measure. The conversion to radian measure is much simpler than the math we'd have to tackle without it. So, rather than think of it as some hard concept they're forcing you to learn, think of it as a clever short cut. That's the way they think of it.

Minima and Maxima

When the catmobile goes over the top of a hill, just for an instant its headlights are perfectly horizontal. That is, its slope is zero. When it reaches the bottom of a valley this happens again. Places where the catmobile changes from going uphill to going downhill (or vice versa) can be quite interesting and useful. They are called "maxima" and "minima." A "maxima" represents the top of a hill, a minima the low spot in a valley. They may be local maxima and minima (the highest summit or lowest valley in a certain stretch of road), there may be several maxima and minima, or we may be interested in the absolute highest or lowest spot on the entire graph. In each case, we're interested in discovering when variables combine to create maximum and minimum situations. What combination of price, advertising dollars, and catnip will create the maximum profit for my new cat toy? When should I leap to reach the highest possible spot on the curtains?

The moments of maximum show up as the crests of hills. One moment the slope is positive, the next it's negative. But for an instant there, right at the top, the slope is zero, our catmobile is horizontal, we have reached a maxima. This spot represents the precise combination of variables that maximizes the variable we're watching. If we're tracking profit, that's a spot we'll watch closely. If we're tracking the likelihood of a dog attacking us, it's the last spot we want to be.

109

On the other hand, if the derivative is negative, then pauses and becomes positive we have reached a minima. If we're watching a stock price, we'd probably want to buy right at that instant, before the stock increases in price again.

Identifying maxima and minima is one of the things calculus does best. Once we have translated our situation into math, then transformed that into a derivative, we "set the derivative equal to zero":

$$\text{derivative} = \text{zero}$$

That is, we create an algebra problem "set to zero" (by putting the derivative on one side of an equals sign and zero on the other) and then solve the algebra problem for the unknown we're interested in. If we've created a function that expresses the moon's fullness as x, and the number of dogs out on the streets as y, we use one of our new calculus procedures to transform this function into its derivative. We are going to be most interested in two combinations of variables: when are there the most dogs on the street (a maxima) and when are there the fewest (a minima). We set our derivative to zero and solve for x.

That tells us when the variables work together to create a maxima or minima, but not which it is. In this case, a graph helps us visualize the situation. We look at spots on the graph to each side of the critical point (when the graph changes direction) and see if we're on a hill or in a valley. If we're on a hill, it's a maxima. If we're in a valley, it's a minima.

A function may have many peaks and valleys, and therefore several maxima and minima, or it may have only one, or it may have none.

The derivative itself is a new function and we can graph it just as well as we could the original. The line we draw may well have minima and maxima on it as well which correspond to the inflection points of the original function.

For the few hundred years of calculus before graphing calculators, students spent hours charting information and creating graphs on paper. This was one of the principal activities of calculus students. As they studied these graphs, they gradually developed an intuitive sense of how various kinds of functions "looked" when transformed into a graph, and how their derivatives looked. They identified maxima and minima on the graph. They also spent hours reading texts which explained the concepts in words. Then they practiced solving many problems with a paper and pencil. If they wanted to know whether something was a maxima or minima they took more than one derivative. Later, the professor would explain it in class and work some problems. Each concept was presented in three or four ways.

Since graphing calculators and computers became common, the calculus learning process has changed. Now it's easy to punch a few buttons and see a lovely curving arc of light which represents something or other. A derivative perhaps? A trig function? We have entrusted the mathematical future of our country to this easy, fun, painless game. We hope that students will see the lovely curving light and understand in an intuitive way what it means.

This is just the way the cats like it.

The Box Problem

We can pretty much guarantee you'll see a problem like this: You've got a flat rectangular sheet of cardboard and you're going to transform it into a litter box. The way you're going to do this is by cutting little squares out of each corner so you can fold the sides up into a box with no top. Your mission will be to create the box that will hold the most litter possible. What size squares do you cut out of each corner?

We don't want to ruin your joy of discovery, but we'll give you a couple of hints. You might consider each square you remove to have a side equal to x. You might recall the formula for the volume of a rectangular box. You might re-read the chapter on minima and maxima.

Second Derivative

We've used the power rule and chain rule and all the rest on a function and now we have the derivative of the function. What happens if we take this new expression, this derivative thing, and use our box of calculus tools on *it*? When we use the power rule and all the rest to differentiate the derivative, we get the "second derivative," a new function which describes how the derivative itself is changing. If our original information is location and time, the function describes our location at some particular time. The first derivative describes the relationship between the two variables and we call it speed. The second derivative describes the relationship between speed (first derivative) and time and we might call it acceleration, for example.

If the headlight on a catmobile cruising over a function-road has traced a line on a big sheet of paper we'd call that the first derivative. If we make a new road which rises and falls to duplicate the line on that paper, our headlights will trace a new shape on the fresh paper before us. This new shape is the second derivative.

Sometimes we know that some combination of variables produces a maxima or minima but we don't know which. We just know that the slope is zero at that spot, but without graphing more than we care to, we can't tell if the spot sits atop a hill or at the bottom of a valley. The second derivative can tell us. If our second derivative is positive, we're in a valley at a minima. If the second derivative is negative, we're on a hill at a maxima. This is called the "second derivative test for maxima and minima." Many people think of the following clown faces to help them remember this:

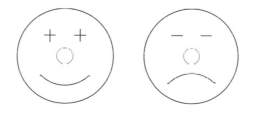

You're probably wondering if we can take the derivative of the second derivative. Yes, of course we can, and the result is called the third derivative. The third derivative is completely useless, and it's a felony to make jokes about them at the airport. Just stay away from them. Even the chocolate ones.

Sequences and Series

Cats obviously love numbers. It's no surprise they like to put a bunch of them together in various ways just to see what happens. It's also no surprise they've given each of them a name.

A sequence is an infinitely long list of numbers that often follows some pattern. We know it's infinite because, after the pattern is established by listing some of them, it's followed by three dots (...) called an ellipsis. The ellipsis means the series just keeps on going without ever reaching an end.

1, 2, 3, 4... is an obvious sequence. 2, 4, 6, 8, 10, 12... is another. These sequences keep getting bigger. That's not always the case. .3, .33, .333, .3333... actually converges. The farther out we extend the sequence, the closer it gets to 1/3.

A series is another long list of numbers. The difference is that a series is a list of numbers that are added together. 1+2+3+4... is a series. Since it's an infinite list, if we actually add them up, we can't ever get a final precise answer. Instead we get a "partial sum." We can add as many of them together as we like to get an increasingly accurate partial sum. Some series also converge:

1+ 1/2 +1/4 + 1/8 + 1/16 ...

This gets bigger and bigger but it never gets larger than some number. For example, it will never reach 2, but you can get partial sums as close to 2 as you like. Your math instructor will say this series has a sum of 2.

Series created from functions rather than numbers are called "power series." Sometimes a power series can be created to solve an unwieldy problem. You may be exposed to them early on because they were involved in the creation of the calculus game, but you won't use them as tools during your first year or so. They are used in engineering and physics, when a close approximation is all that's needed. They are also useful because they allow us to use simple arithmetic operations, like adding and subtracting, to calculate values of transcendental functions, like trig functions and root extraction.

Asymptotes

When graphing functions in the xy plane, it sometimes happens that, as the graph heads off the page, it becomes more and more close to a particular straight line. That line is called an asymptote (ass-em-tote).

Exponential Functions

In an exponential function something is raised to a *variable* power.

Something raised to a *fixed* power is *not* an exponential function. To qualify, the variable has to be the exponent. For example:

$$2^x$$

is an exponential function. Two is the base, x (the variable) is the exponent.

On the other hand, in this:

$$x^2$$

x is raised to the power of two; x is the base and 2 is the exponent. Although we may enjoy playing our algebraic games on this function and using our knowledge of exponents in general, it is *not* an exponential function. The exponent isn't a variable.

Exponents in general are useful critters because we can use what we know about them to devise shortcuts. Rather than multiply

x^2 times x^3 we can simply add the exponents and come up with x^5, for example. Each time we translate a difficult math problem into something that can be solved with the little tricks of exponents, we save time and brain damage.

Calculus takes this a step farther. The derivative of an exponential function is always proportional to the original function; it's the function times a constant. Even better, if we can translate a thorny problem into a special kind of exponential function, the answer rolls off the page and into our laps. The constant is one.

Exponents are abbreviations for the process of multiplying a number times itself then multiplying the answer by the original number again, then multiplying that result by the original number over and over again. The number you start with is called the base. The number of times you use the base as a multiplier is the exponent and is designated by a small number raised slightly to the right of the base. In

$$2^4$$

2 is the base and 4 is the exponent. You will use 2 as a multiplier four times:

$$(2)(2)(2)(2)$$

If you did the math you'd learn this is the same as the number 16.

Logarithms are the reverse process. They start with the 16 and say "To what power must I raise a base of 2 to get 16?" The answer, of course, is 4. In math class, it will look like this:

$$\log_2(16)=4$$

Until computers and calculators became common in the last half of the twentieth century, if you loved math you loved logarithms. Mathematicians made charts of the logs of many numbers and used

116

them to solve tedious problems quickly. Rather than multiply huge numbers times each other, for example, they simply looked at their charts, converted the numbers to logs, then added those two numbers together. After a simple conversion back to real numbers they had their answer. It took about an inch of scratch paper rather than three pages of computations. Even considering the conversions, adding is easier than multiplying.

Now, rather than use charts, these calculations have been built into your calculator. But the concepts remain important to solving calculus problems.

When we chart things like population growth, radioactive decay and continuous interest on our money, then convert the chart into a function, we are likely to find an unknown like x as an exponent. We'll have an exponential function. Exponential functions usually involve time.

In an amazing twist of fate, the derivative of an exponential function is that same exponential function times a constant.

$$y = 23^x$$
$$y' = (\ln 23) 23^x$$

The only trick is figuring out what that constant is going to be. Sounds easy, right?

But not easy enough to please a cat. What a true cat would require is some way to make sure the constant was always a one. That way, the derivative would always be the exponential function you started with, times one, which is dog-simple math. Another way to look at the same idea is this: is there some actual number you can raise to any power and its derivative will be the same function you started

117

with? Is there some number you could use as the base of an exponential function and its derivative would always be an identical exponential function? If there was, and you simply converted your trickier problems into exponential functions with this number as the base, your life would be a lot easier. Is there some number like that?

Yes there is.

But it's not a garden variety number you would instinctively multiply times itself a bunch of times. In fact, it's impossible to accurately multiply it times itself even once. This does not seem to bother mathematicians. By the time they start thinking about things like this they have wandered so far from the flight path of everyday instincts most of us can no longer even see their tail lights winking in the night. The number is an irrational number, like pi, which results from trying to solve this little problem:

$$1 + 1 + \frac{1}{2!} + \frac{1}{3!} + \frac{1}{4!} + \frac{1}{5!} \ldots = ?? \ \dot{=} \ e$$

The result is the number 2.71828... In honor of a German cat named Euler (pronounced "oiler") this number has been given the name e. Actually, Euler gave it the name "e" but modestly insisted it was short for "exponent" not "Euler." Whenever you see this little e in italics in a math class, they are abbreviating 2.71828... If $y = e^x$, the derivative is $y' = e^x$

This is obviously way too cool to ignore. If you haven't already taken too many math classes you will step back for a moment, marvel that there is such a number, then ask "why does this work?"

Remember that finding the logarithm is the reverse process of raising to a power, like squaring and taking the square root are reverse processes. When we raise two to the power of two (2^2) we get four.

118

$$2^2 x = 4^x$$

When we take the square root of four we get two. When we raise 10 to the fourth power, we get 10,000. When we find the log of 10,000, we get 4.

If we convert an exponential function into its equivalent with a base of e, the calculus part of the problem is simple. The idea of multiplying e times itself may seem very abstract and weird, but we don't care about the calculating because we probably won't ever do it. If it's possible to simply translate an exponential function into an equivalent function expressed with e as the base rather than some more normal number, we'll do all our calculus then translate back without doing anything to the numbers while they are temporarily base e creatures. We'll disguise our geeky function as a football player function long enough for it to run across the field and safely reach the correct dressing room, but we don't really care that it can't catch a football. We just won't throw one its way. Once it's back where it belongs, it'll change back to geeky street clothes.

We call the logarithm with base e the "natural log" which is often abbreviated "ln". We know that, by itself, $\ln(e)=1$, and our calculators can figure what the natural log of other numbers is without us having any idea what it means to take the natural log of something other than e.

Obviously we'll want to transform ugly functions into this elegant format.

It turns out we can use the same rules of exponents use learned in *Algebra Unplugged* (a book written by the authors of this book) to do the trick. Remember that

$$x^{2+3} \text{ is the same as } x^5.$$

That's one little trick of exponents. And remember that

$$2^{2x} \text{ is the same as } 4^x$$

$\ln(e)=1$ Base $e = \ln = $ NATURAL LOG

Wait a minute! Notice what happened in that last example. The answer is the same, the two expressions are equivalent. But in 2^{2x} the base is 2, while in 4^x the base is 4. Simply multiplying a constant times the exponent effectively changes the base. We've used our extensive knowledge of how exponents work to change the base.

Less obvious but just as true

$$e^{(\ln 2)x} \text{ is the same as } 2^x$$

We can use the easy principles of exponents to convert an expression or function so that its base is e and therefore under the spell it casts on derivatives. The procedure itself is simple: we change the base to e, then multiply the exponent by the natural log of the old base. If we start with 2^x we do those two things and our result is:

$$e^{(\ln 2)x}.$$

We have converted our exponential function into an exponential function with a base of e.

If you can manipulate a function in this way and convert it into an exponential function with e as the base, finding the derivative is child's play. The derivative is identical to the function.

This number e is extremely useful when we study rates of growth or decay, things that translate well into exponential functions. If you convert the exponential function that describes the growth of your savings account into this world, and you are lucky enough that the new function is e^x, then at the end of one year the $100 you started with will have grown to $271.825. If you know that your money doubled in one year, you could figure the rate of growth as .69. If this doesn't make much sense to you, that's okay. It's the result of figuring a 100% rate per year according to the standard formula which says that "the amount in your account (A) equals your the original principal (P) plus interest." It looks like this:

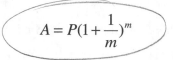

$$A = P(1 + \frac{1}{m})^m$$

which gets closer and closer to e as you compound your interest more and more frequently. Here, m is the number of times you compound per year.

In other words, if the rate of interest is 100%, and the interest is compounded annually, and you start with $100, at the end of a year you'll have $200. But if the interest is compounded continuously (that is, every instant), at the end of a year you'll have $271.83. This mysterious e represents the effect of continuous growth, or continuous compounding, on a fixed rate.

$$e(\ln 2)x = 2^x$$

$$2^{2x} = 4^x$$

In the early 1900s, steam engines and internal combustion engines were equally popular. Steam cars had many advantages; they were more powerful, faster, simpler, and didn't pollute. Their only drawback was you had to stop and add water too often. When Henry Ford decided to use the techniques of mass production that had worked so well for gun makers, he nearly built a steam car. Finally, he chose the internal combustion engine for his Model T and left everyone else in his dust. He knew that steam engines were superior in many ways, but decided that internal combustion engines would be easier to build using the manufacturing techniques of the day. The fact that internal combustion engines are noisy and pollute wasn't an issue at the time; no one cared. It didn't matter that steam engines could run on natural gas or grain alcohol. We had plenty of oil.

Steam engines disappeared not because they were inferior but because the Model T made so much money that other car companies, scientists, engineers and consumers followed along behind the trail it blazed. Inventing improvements on the internal combustion engine could make you rich, inventing improvements on the steam engine just seemed weird.

If Henry Ford had picked the steam engine, the exact reverse would be true today. We'd all be driving silent, pollution-free steam cars. Things like fuel injection systems and catalytic converters would never have been invented, because they were created to solve problems the steam engine doesn't have to begin with.

In the 1960's there were two types of computers, analog and digital. Although analog computers had the advantage of using the

[handwritten note in left margin: Made oil companies Rich!]

122

kind of "fuzzy logic" used in human thought, they were abandoned when someone realized they could make lots of money by using the digital approach to build hand-held calculators. Analog computers disappeared. Today the only folks building analog computers are guys who use them to control little robots. Had the analog guys figured out a way to make a bunch of money early in the game, all our computers would use a completely different approach, which would have its own advantages and disadvantages.

The history of math is full of similar choices. Someone wrestling with a nasty problem would come up with a new strategy and everyone else followed along. Alternative strategies were abandoned as the math community surged down an intellectual fork in a road.

Once you get far enough down any path, it becomes hard to go back. This is why we don't have steam-powered lawn mowers today. They would be powerful, quiet, cheap and non polluting. We could add water with the garden hose when we need to, eliminating the one built-in drawback. The only disadvantage is that someone would have to return mentally to the early 1900s and start up where Henry Ford and his buddies left off. So far, that hasn't happened.

All the choices mathematicians have made over the centuries created the finely tuned machine that math is today. These choices have also left us with some weird quirks, but it's unlikely the entire math battalion will casually go back and choose a different fork in the road. Students just have to accept "the way it is" while they're learning. Of course, once you get the hang of it, there's nothing to prevent you going back and trying some different fork in one of the roads. It's easier to forgive mathematicians for some aspects of math if you understand that no one came up with the entire system. They inherited a complex but imperfect machine and used coat-hanger wires and duct tape to improve it when necessary so it could keep up with new problems to solve. Learning math is really learning a system based on all these choices, not necessarily learning one true science. Although some folks disagree, it's easy to believe that math is an invention, not a discovery.

Back in the days when math was used primarily for counting the number of wolves attacking us and the number of arrows in our quiver, intellectuals of the day decided that numbers should be units of uniform size. Adding one to a number should increase it the same amount whether we're adding one to six or one to fifty. A big wolf and small wolf are equal for the purposes of counting them. This has become so ingrained over the last few thousand years we can't imagine any other way to think about it. But if we had begun inventing numbers in a world where our major problems were circles and diameters, rather than wolves, we might have chosen differently. In that reality, pi would be a simple rational number.

The Babylonians divided a circle into 360 parts giving us our modern degrees. This probably had something to do with their 360-day-per-year calendar, with a five day party at the end. Had they divided it into 500 parts, that change would have trickled down as well. The British came up with "feet" as a rough equivalent of one large man's foot. The French came up with the "meter" by dividing the distance from the North Pole to the equator into equal parts. Now we have to live with both units of measurement. We chose the number 10 to build our number system around. Other cultures did not. There are thousands of examples. They don't represent "right" or "wrong," but choices made years ago.

The Greeks held geometry to be sacred in almost a religious way. Pythagoras belonged to a secret math society devoted to geometry; members were killed for revealing its secrets. If you were Greek, all math had to be expressed in the language of geometry, with its rigid rules and proofs. Certainly this idea informed Descartes when he started plotting lines with x and y coordinates. It continues today as calculus instructors demonstrate how functions interact in a graphic way. And the Greeks have wormed their way into my brain and yours when it comes to learning math. We want a geometric analogy to every new concept.

Geometry works like a charm when lines are straight and functions are steady. Curved lines require fancy arithmetic and our results don't always fit nicely into the even older tradition of uniform units. Since we weren't on the committee that chose which duct tape to use to keep the thing running, it may not seem intuitive or even logical to us. We may discover ancient Greeks arguing loudly with medieval German mathematicians in our brains. The cats love this.

The idea of multiplying a number times itself seems harmless enough. But cats can take the simplest idea and tangle it into a nightmare of mathematical games drawing the unsuspecting human deeper and deeper until they lose sight of the distant shores of reality.

When you multiply a number times itself you call it "squaring." Utter the word and immediately a chorus of ancient Greek mathematicians trots on stage to remind you that this is exactly what happens when you create a geometric square, with each side representing the number you're multiplying. Five squared, they intone, means a geometric square with each side five units long. Behold! The area of this square is 25 square units. The Greek mathematicians smile and nod to each other in satisfaction. We nod with them and think of the area of anything as a certain number of squares.

If you repeat the process (five times five times five) you write it 5^3, which is pronounced "5 cubed." The Greek chorus is quick to remind you that this is the same as creating a cube where each side equals five. The volume is 125 cubic units. We nod again and agree that this makes sense. The volume of the cylinders in our motorcycle might total 750 cubic centimeters

In a puckish mood you decide to tweak their collective Greek tails, so to speak, and inquire about raising it to the next level: 5^4.

This is called "being mean to Greek mathematicians."

There isn't a geometric model for raising something to the fourth or fifth or any other higher power. Geometry, so useful in its place, falls on its face when you try to extend the analogy. But there they are, Greeks in our brains, screaming for a geometric explanation of exponents that doesn't exist and so we become confused and frightened.

If we can ignore the Greeks, the concept of exponents is quite useful. When people began using exponents they thought of them as "multiplying a number times itself repeatedly." Soon they noticed interesting characteristics which allowed them to create new exponent games in which they never got around to actually "multiplying a number times itself." In fact, they created games in which that would be impossible to do and get a precise answer.

We can translate the basic structure of our traditional "base 10" number system into a system of exponents. One is the same as 10 raised to the zero power—anything in the one's column is the same as that number times 10^0. The ten's column is 10 to the first power, the hundred's column is 10 to the 2nd power, etc.

1	10^0
10	10^1
100	10^2
1,000	10^3

It didn't take long to notice that multiplying two numbers got you the same result as adding their exponents from the right column. Ten times one hundred is the same as 10^3.

No cat could resist such elegance. Before calculators were invented, multiplying big numbers was a long and tedious process, and it was easy to make mistakes. Adding was much easier. Even in our example, adding one plus two is easier than multiplying 10 times 100. As the factors become more complicated, the advantage becomes more pronounced.

It occurred to a fellow named Napier that the world would be a more pleasant place if you could convert any number into 10 raised to some power. Then you could perform the most complicated multiplication problems with one simple addition. Napier happened to live the same time as Galileo and Kepler, who were compiling lots of astronomical data. To interpret this data they were faced with routinely multiplying huge numbers, so there was an immediate practical benefit to his idea.

To accomplish his plan, he extended the idea of exponents to a much less intuitive region, one the Greeks would not have enjoyed at all. He assumed you could multiply a number times itself half a time, or 2.353 times, or any other fractional part of a number. As long as the rules of exponents survived, we could raise a number to any power we wanted, even to the power of an irrational number like pi. This made possible a whole new way of performing multiplication on large (and weird) numbers. To multiply $10^{2.5}$ times $10^{2.5}$ we never have to actually perform the exponent part of the problem. We simply add the exponents, and the answer is 10^5.

Figuring out exactly what exponent to use for each number required long and tedious calculations, but Napier took on the challenge. He spent 20 years creating tables of these numbers. His new system was a big hit, and the tables he and his collaborators made were used for 350 years until calculators were developed and multiplication of large numbers became easy.

Like most good math, his idea involved two processes that were each the inverse of the other. That is, one reversed the other just as addition reverses subtraction or squaring reverses taking the square root of a number. It gets you back where you started.

The reverse of raising 10 to a power is called "finding the logarithm of the number." If you raise 10 to the second power, you get 100. The logarithm of 100 is two. When you are asked to find the log of a number, your answer will be the power you raise 10 to get that number. The answer is the exponent.

Because we adopted a "base 10" system a couple thousand years ago, and because of all Mr. Napier's tedious work (and the work of a fellow named Briggs, among others), when we search for logarithms, we all agree that 10 is the number we're going to raise to some power. It's the base. We don't even bother to write it in the question; we say "log 100 =?" and we know the question is another way of writing $10^? = 100$. The answer is two.

Logarithms are used today in the Richter scale for measuring earthquake intensity, the pH scale for measuring acidity, and the decibel system for measuring loudness.

Because of Napier's work we could find the log of pretty much any number by looking it up in his chart and using his relatively easy

128

process for completing the multiplication. To multiply 67,876 times 99,767 you use his chart to find the log of each one. Because these logs are exponents of 10 you simply added them, then converted the result back to regular numbers. In about thirty seconds you could complete a multiplication problem that might have taken you 20 minutes otherwise. As LaPlace said, the invention of logarithms "by shortening the labors doubled the lives of astronomers."

Of course, you can take numbers besides 10 and raise them to whatever power you want. Napier didn't create charts for the other numbers because base 10 had already become the standard. But, had their been a good reason to do so, and if Napier lived an inhumanly long life, he could have.

Just as raising something to a power is a repeated multiplication problem, taking a log is a repeated division problem. The log of 8 base 2 is 3 because $8 \div 2 \div 2 \div 2 = 1$. We used the base as a divisor 3 times. As we continue to divide, at some point our answers become smaller than the base itself. We can't always divide every base by every divisor an even number of times and reach 1. Decimals were invented to describe the results of these repeated divisions.

It turns out there's a reason to make a chart for one number besides 10: the irrational number 2.718281828... also known as "e". The reason is that the derivative of e^x is e^x. The derivative of e^x when $x=275$ is e^{275}. Again, this was way too cool to ignore.

It is intuitively odd to multiply an irrational number by itself because you'll never be precisely accurate. Squaring 2.7 results in a slightly different answer than squaring 2.71, which is slightly different from squaring 2.718. Luckily, we skip that part of the problem.

Finding the Derivative of Exponential Functions

To find the derivative of any exponential function, we convert it into an exponential function with a base of e rather than 10, find the derivative (using the chain rule nearly every time), and convert it back to base 10 before anyone notices what we're up to. It's not even difficult math; we just use the rules of exponents someone developed when they ran out of duct tape. We change the base to e, then multiply the exponent by the natural log of the old base.

The Greeks would have a fit. We're taking a number we can never perfectly represent geometrically, and pretending to multiply it times itself some number of times, and the number of times may itself be an irrational number. Hey, Euclid, scratch a shape in the sand that represents that! Eureka my foot!

A calm observer might respond that e, and the way we use it, are simply quirks of all the choices made by mathematicians through the ages. They started by creating numbers, then arithmetic, then geometry and algebra. By the time they got to calculus, all that was well established. If they'd started out inventing math by focusing on exponential functions and the first beast they created was not "one" and the other numbers, but a function you can raise to any power and it would be its own derivative, and if they based the rest of math on that, calculus would be a breeze. On the other hand, you'd probably need a super computer to count mice in your living room.

The Greeks were obviously not "cat people." For the price of a saucer of milk they could have gotten a nice geometric model. You will not see the following analogy in any other calculus book; we made it up because we have very loud Greeks clamoring in our own heads. It may be more confusing than illuminating. If you hunger for a geometric model of what happens when you convert to base *e*, it may satisfy your craving. If it's not useful to you, forget it. It won't be on the test.

A cat named Origin stands in the center of a frozen lake holding a rope. A second cat named Loopy holds the other end and skates around him. Centrifugal force keeps the rope taut. If Origin doesn't let out any rope, but simply rotates in place, Loopy will skate a perfect circle around him.

If Origin slowly lets out rope, Loopy will skate in an ever-expanding spiral around him.

The lake is covered with a thin layer of new snow so Loopy's skates leave a trail. This trail is a record of Loopy's path. You could say it is a graph of several variables: how long the rope is at any moment, what direction Loopy was skating at any moment, and how far has Loopy skated altogether. If we're good detectives, we can probably figure out a lot of information from the clues provided by that spiral in the snow.

Of course, Origin doesn't let the rope slide through his paws in some random fashion. Like any good cat, he is guided by a function. Different functions cause him to release rope at different rates. Loopy's path graphs the ever-changing distance between him and Origin over time. You could say it graphs Origin's function.

The angle the rope in Loopy's paws makes with his track in the snow is related to the rate of change in the length of the rope at that instant. The faster Origin plays out rope, the faster Loopy will be moving away from him, and the greater the angle of the rope in Loopy's paws. Alternatively, Loopy could keep the angle steady and skate while Origin merely plays out rope to keep it taut. In this case, we know the rate of growth and we're interested in watching how that rate affects the way Origin lets out rope to accommodate it. We know the rate of growth and we're interested in how much something actually grows.

If Origin obeys the commands of some exponential function, the rope in Loopy's paw will make a fixed angle with his track in the snow. Or, if Loopy controls the release of the rope, and holds his arm at a constant angle, Origin will be forced to release rope according to some exponential function.

Perhaps the cats are interested in graphing their herd of mice with this strange rope and skating graph. They have 100 mice in the herd now, so they let out the 100 feet of rope and Loopy skates a circle around Origin. That represents no change in the size of the herd; the rope still stretches 100 feet between them. The rope makes a 90 degree angle with the track in the snow. Or, as you might want to say, the rope makes a 90 degree angle with a line that is tangent to the track in the snow at its point of intersection.

Of course, part of the joy of raising mice is watching your herd multiply and grow. This is exactly the kind of process our spiral graphs best. Mice typically reproduce at a rate of 100 percent per year. That rate is represented by how fast Origin plays out rope. He releases rope so that exactly twice as much rope is stretched out after one complete rotation as he started with. One complete rotation represents the time period given in our rate. Mice are traditionally counted once a year, so one revolution equals one year. Had we been interested in miles per hour, one revolution would have equaled one hour.

Loopy can keep skating all day long, with Origin playing out rope as the original function dictates (doubling the length every rotation in this case) to determine the size of the herd in five years or 10, or any other length of time.

A growth rate is easy to describe in terms of some specific length of time, ("the population doubles in one year") but describing the way mice actually multiply requires a more subtle function. Every day a few mice reach reproductive age and start to multiply, then their babies grow up and start to multiply as well. To represent this compounding of the herd we need to use an exponential function. Time is one issue, the size of the original herd is another, and so is the rate of reproduction. But we also have to figure in the compounding effect of babies reproducing. This is similar to adding interest to our savings account's principle balance and earning interest on the new, larger amount. The specific number of new baby mice does not increase in a steady way. You may count 10 new babies on August 1, but you'll probably count more on October 1, and many more on December 15.

The magic decoder ring that lets us express this compounding is our old friend e. It represents the change between a stated rate of growth (like "10 percent per year") and how much something actually grows over time if the growth is compounded. As we compound more frequently and examine smaller increments of time, and therefore smaller amounts of growth, we keep discovering e in our calculations. The shorter the increment of time, the more decimal places we learn in e. If it hadn't been for the Egyptians and Babylonians and Greeks and Germans who all contributed choices to our mathematics, this might be represented by some gentler looking beast. As it is, if we want to keep pi, and trigonometry, and tape measures, we need to accept that e is the number we need to express this relationship, just as we need to accept pi as the relationship between a diameter and circumference of a circle.

When we have done dozens of problems with different interest rates and growth rates we become more and more convinced that *e* has a real place in our system. No matter what growth rate you use, or what period of time you use, as you examine the growth of an exponential function over a tiny period of time, it will always appear. As you consider shorter and shorter periods of time in search of the instantaneous rate of growth, you'll discover that *e* appears with more and more decimal places. Since our real-world concept of rates involves two measurements some distance apart, we won't ever reach a single number that represents it. This shadowy *e* character lets us approximate an instantaneous rate (a rate with only one measurement) without creating a situation where we have to divide the distance traveled (or the amount of growth) by zero time. If we came up with a rate having a zero in the denominator, we couldn't use algebra to finish the problem because we can't divide by zero. This is a piece of duct tape added centuries ago to allow all the rest of algebra to work and it's dear to the hearts of mathematicians everywhere. Had ancient mathematicians used coat hangers or glue to get around that problem, *e* might not be so important now.

The particular spiral that describes exponential growth is called a logarithmic spiral. It has several interesting properties and was once considered magical. If Origin plays out rope according to an exponential function, the angle of the rope in Loopy's paws won't change compared to his track in the snow. Or, if Loopy holds the angle of the rope steady, Origin will be forced to play out rope according to an exponential function. This is a nifty geometric illustration of exponential growth. Also interesting is that a straight rope from the origin crosses all the tracks in the snow at an identical angle.

If Loopy skates around Origin and the frozen rope makes a 45 degree angle with the track in the snow, he's plotting an exponential function with *e* in the base. That is, the length of the rope is equal to *e* raised to a power. If Loopy knows how far around Origin he's skated, he can use the radian measure of that angle as the exponent of e and calculate his exact distance from Origin. This rarely happens naturally, of course. Usually Origin plays out rope according to an exponential function and Loopy's arm is stuck out at some angle besides 45 degrees, or Loopy holds his arm at some angle and Origin plays out rope accordingly.

But if Loopy's arm is stuck out at a constant 45 degree angle, and if you express Loopy's path in terms of radian measure, and the function that describes the rate of change of the length of the rope at any instant is identical to Origin's "play out the rope" function at that instant. In your Required Calculus Textbook, they will say it something like this: "If x tells the radian measure of how far around Loopy has skated, the derivative of e^x is e^x."

Perhaps the situation calls for a specific rate of growth. Loopy holds his arm at the appropriate angle, Origin plays out rope as Loopy skates around him. Like most cats, Origin has a photographic memory for functions and memorizes how he plays out rope. When they're done, they decide to convert the exponential function they just graphed to a base e exponential function. Loopy sets his arm at 45 degrees and skates while Origin plays out rope exactly as he did a moment ago. Loopy may have to skate faster or slower, but we don't really care. What will be different is that before one full rotation was one year, but now it may represent some other time period. The new tracks in the snow contain all the information of the original function, but they make a new path in the snow. This new track represents the original exponential function converted to an exponential function with a base of e. In effect, Loopy and Origin have physically modeled the equation $a^x = e^{(\ln a)x}$.

Perhaps now the Greeks within our brains will shut up.

$$a^x = e^{(\ln a)x}$$

$$x^{-1} = 1 \div x$$

Another Reason They Like e

Mathematicians love consistency, and hate it when their games contain even one little spot that doesn't seem to fit. They love it that the rules of manipulating exponents fit right in with the rules for finding derivatives, for example. But when they made a chart of all the various powers you can raise x to, and set that next to the chart of their derivatives, there was a gap. Every x raised to every power was the derivative of some function except for this one critter, x^{-1} (which is the same as $1 \div x$). No power of x gets that as a derivative, no matter how good they are.

This wouldn't bother you or me so much. We're used to the little inconsistencies of life and accept them. Sometimes the cat sits on our lap, sometimes it treats us like a stranger. It doesn't mean they're responding to commands from some distant planet or anything, right? But it bothers some of the math guys.

Luckily, once the idea of exponential functions with a base of e occurred to them, and along with it the idea of the natural logarithm, suddenly x^{-1} fit right in. By adopting e, this annoying little inconsistent gap vanished. The derivative of $\ln x$ is $\frac{1}{x}$. The mathematicians would have embraced e and natural logs had they done nothing more than this. It allowed them to sleep at night.

Strategically, by using the natural log and e, we expand our bag of tricks. We could translate functions into other bases, but they don't convey this strategic advantage. In every case but one, the anti-derivative (that is, the original function) of a derivative x to a power is, roughly, x to another (one higher) power. That pattern fails in a big way for x^{-1}. Mathematicians and cats perk up their ears when a pattern fails. What would the anti derivative be in this case? It's the natural logarithm function.

The exponential function e^x is the answer to a second question that every calculus instructor thinks is cool: Is there a function that equals its own derivative?

One Common Use of e

Some exponential functions are so useful in a particular field, like biology or business, everyone in the field just memorizes the function and uses it over and over. One example is how money grows when compounded continuously, which is identical to how populations grow. The letters change to reflect the field of study, but the formula remains the same. In business, the amount of money in your savings account is abbreviated with a capital A, the rate of interest is abbreviated r, (or i, which is interest per period of time), time is t, principal is a capital P, and the total number of compounding periods is abbreviated with an n. The formula for figuring out the amount of money in your account looks like this:

$$A = P(1+i)^n$$

With the limit process, you might describe the continuous compounding of your savings account like this (the horizontal figure eight stands for infinity.):

$$\lim_{n\to\infty}(1+\frac{1}{n}) = e$$

After lots of fun math, the original formula gets simplified to

$$A = Pe^{rt}$$

That's easy to remember, because you could pronounce the right side of the equals sign "pert," which many people do. If you want

$A = P(1+i)^n$

to know the amount of interest in your account at any time, and your money is being compounded continuously, multiply your principal times *e*, raised to the power of "the rate times the time", and you've got your answer. If your savings account is like mine, you'll never need calculus to figure its balance. But if you're computing how many bacteria are in the swamp, you'll love this little short cut.

With Respect to Time

When we plot a graph involving two variables we often know the information on the horizontal axis and are curious about what the vertical component is at a particular point. After a certain number of hours in a math class, we tend to think of the known variable as x and the unknown as y. When we figure the relationship between the two, or the derivative, it will be a fraction with the y component on top. y will change a certain amount for each change in x. That is, we are finding the slope with respect to the horizontal axis, or finding the derivative with respect to x.

Often we are interested in how one or more functions behave with respect to time. For each second that ticks by, how much will y change? When this is our interest, we take derivatives with respect to time. That is, we no longer consider the horizontal axis to be x, or some unknown specific to our problem. We consider the horizontal axis to be t, for time. Our derivatives will by dy over dt.

This is especially useful when we have more than one function operating on something. It may be difficult to describe how one variable affects a process compared directly to another variable. If we can compare them each to time we can relate the two variables to each other. It's a little like using common denominators to add fractions together.

The phrase "with respect to" tells us what we're using as our known quantity or independent variable. It tells us what the horizontal axis represents. It tells us what will be on the bottom of a derivative expressed as *dy* over *d (independent variable)*.

Parametric Equations

Sometimes it's hard to compare two things directly. It's hard to compare Clark Kent's height to Superman's height directly, for example, because you can't ever get them into the same room at the same time. But you can measure Clark against Jimmy, then Superman against Jimmy, and pretty soon you figure out all you need to know.

When we use one variable as a standard (or parameter) to measure two or more things against, we call it a parametric equation. The most common parameter to measure other things against is time. We can figure where Superman will be at noon, then where Jimmy will be at noon, and where the bad guys will be at noon. Obviously this is most useful to us if the things we're measuring are related in some way. In math class, this is usually accomplished by putting each of them on a different train with different insane cats at the controls.

If we compare *x* to *y* we get this:

time (*t*)	*x*	*y*	relationship
1	2	6	$y=3x$
2	4	12	$y=3x$
3	6	18	$y=3x$
4	???	24	$y=3x$

$$dy/dx$$

When y is thought of as a function of x, the derivative of y with respect to x or dy/dx is 3. For each unit of change in x, y changes by 3.

If we compare x to time we get this:

time (t)	x	y	relationship
1	2	6	$x=2t$
2	4	12	$x=2t$
3	6	18	$x=2t$
4	???	24	$x=2t$

There is more information here because both x and y are described as functions of t. When we think of x depending on t, we use the formula $x=2t$. Its derivative with respect to time, dx/dt is 2. In each second, x goes up by 2.

If we compare y to time we get this:

time (t)	x	y	relationship
1	2	6	$y=6t$
2	4	12	$y=6t$
3	6	18	$y=6t$
4	???	24	$y=6t$

When we think of y as a function of t, we get $y = 6t$. Now, the derivative of y with respect to t, dy/dt equals 6. In each second, y goes up by 3.

It's no coincidence that 6 divided by 2 equals 3. In derivative notation, we can write a relationship that almost always is true when variables are related parametrically:

$$\frac{dy/dt}{dx/dt} = \frac{dy}{dx}$$

Related Rates

We can describe some real world situations different ways depending on which aspect of it we're interested in at the moment. We might describe a house in terms of the square footage of the floor if we're buying carpet, but describe it in terms of its volume if we're choosing a furnace, or the surface area of its walls if we're buying paint. Instinctively we realize that these descriptions are all related in some way. A bigger house will require more paint, more carpet and a larger furnace than a small house. When we describe each in mathematical terms, we'll come up with three different formulas. These formulas are also related in some way.

Once we decide on the house's dimensions, we can extract the necessary information with relatively simple math. But while we're making plans, it may be useful to learn how each changes if we change one. For each foot taller we make the house, how much larger a furnace will we need? We might even use the cost of materials as part of our formulas to see whether it's more cost effective to build a two-story house, or a one story, or if a long skinny house is smarter than a square house.

We could do this by drawing a hundred different house plans, figure the components for each and when we're done, days later, compare them.

Or we could use calculus to describe how each changes as the others change and answer our question with a single calculation. Problems of this sort are called related rates problems.

In your first semester of calculus you'll become intimately familiar with two types of problems: maxima/minima problems and related rates problems. Related rates problems usually involve things that change over time and things that are growing, shrinking, accelerating or otherwise changing from moment to moment. There are five steps to solving a related rates problem.

Step One

Identify the rates of change involved in your problem.

Step Two

Draw upon your massive knowledge of various standard formulas and identify some formula that contains elements common to these rates. Perhaps this will be the formula for volume of a cube, or the area of a square, or the Pythagorean Theorem. Obviously, you'll want to pick one that is meaningful in the context of the problem.

Step Three

Use that formula to create an equation that includes all the variables you're interested in (perhaps length and volume). You may have to be creative and modify it in some way so that your equation doesn't contain any extra unknowns.

Step Four

Take the derivative of both sides of your equation with respect to time.

Step Five

Go back to the original problem, plug in the information you were given and solve for the unknown.

The standard mistake is to try to substitute some known information or algebraic equivalent too soon. You want to leave the formulas fluid until you've found the derivative. Until then, you're interested in the functions and how they interact, not the specifics that were given to you. At the point you're interested in, if $x^2=4$, you might be tempted to replace *x* with 2. But once you do that you've taken a snapshot of a particular instant rather than letting the squaring function survive to the derivative stage.

You will know that you're embarking on a "related rates" adventure this way: you'll be given rate information and you'll be asked to find rate information.

Cubic Mice

You've never seen a cubic mouse, but they are loathsome and curious mythological creatures. Because of a genetic quirk, cubic mouse are shaped like perfect little cubes covered with nasty brown fur. Unlike most mice, all the food they eat is converted into body mass, so they grow large quickly. Because of this, and because they are so easy to stack for storage, this is the variety most preferred by cat ranchers who raise them as humans raise cattle.

If the cats feed them a cubic inch of grain each day, they will grow exactly one cubic inch in volume. Obviously, cubic mice feature prominently in agricultural courses for cats. Cats often use related rates techniques in their herd management.

For example, a cat may be given a problem like this: A cubic mouse is growing three cubic inches per day. When it is 64 cubic feet big, how fast is each side growing? They choose how to describe the situation and decide *x* = the length of one side, *v* = volume, and *t* = time.

They turn to the five steps of solving related rates problems.

143

Step One

What are the rates we're interested in? One is the rate the volume is growing. The answer is three cubic inches per day or:

$$dv/dt = +3$$

The other rate is the growth of a side, which is:

$$dx/dt = ???$$

Step Two

They search for a formula that has both x's and v's (one side of the cube and the volume of the cube) They know that volume of a cube equals one side cubed or:

$$V = x^3$$

space

Step Three

They create an equation that contains both variables, that is, both x's and v's. Luckily, the volume formula already does this, so they write it down:

$$V = x^3$$

now plug in time

Step Four

They take the derivative of both sides with respect to time:

$$dv/dt = 3x^2 \, (dx/dt)$$

This is the equation that relates the rates to each other. They know that $dv/dt=3$ and need to find dv/dt.

Step Five

They go back to the original problem and plug in the information, then solve for the unknown. The known value is that the volume is 64, x is the unknown side, dx/dt is the rate a side is growing when v=64.

144

What is x when the volume is 64? It's 4. This is the spot they're interested in. At this point, they substitute 4 for x, and the derivative they've got nailed down:

$$3 = 3(4)^2(\frac{dx}{dt})$$

Then they solve for dx/dt:

$$3 = 48(dx/dt)$$

or

$$\frac{3}{48} = \frac{dx}{dt}$$

Then they puzzle over what this means and decide that when the mice are 64 cubic inches big they are growing at a rate of 3 inches per side every 48 days, or 1/16" per day.

This is the answer to the question: "how fast is each side growing when the mice are 64 cubic inches big?"

Partial Derivatives

If a problem only has two variables, we feel pretty comfortable, but many situations require three or more variables to describe them. The formula for the volume of a regular old cylinder, like a can of cat food, is like that:

$$v = \pi r^2\, y$$

Which translates to "volume equals π times the radius squared times the length of the cylinder."

The problem is that a change to either the radius or the length will affect the volume. We've got two independent variables. The way we get around that is to assume that one doesn't move while we're looking at the other one. We pretend it's a constant and figure the derivative. Then, if we need to, we switch and, using the same trick, figure the derivative with respect to the other variable.

In more general language, when we do this, we call volume z, use x for r, and y is still y. So now it looks like this:

$$z=f\,(x,y)=x^2y\pi$$

Constants are easy to deal with in calculus, so doing this lets us concentrate on one part of the problem without adding a lot of extra work.

Inverse Functions

[handwritten: $y = x^2$ on $x = \sqrt{y}$]

In a certain situation we can think of y as a function of x or we can think of x as a function of y. Perhaps one side of a square is x and the area is y. We can think of the situation as either $y = x^2$, or we can think of it as $x = \sqrt{y}$. These two are inverse functions. There is a short cut for finding the derivative of one if you know the other:

The derivative of the inverse function is the reciprocal of the derivative of the direct function.

If you come to love graphs the way most math instructors do, you'll be interested to learn that inverse functions make mirror image graphs.

146

If a problem comes out and tells you that x is a function of y like this–

$$x=y^2$$

–we can find the derivative of the function of x with respect to y, which will look like $dx/dy=2y$. Perhaps a square cat box is y inches per side, and has an area of x. In this example, it's obvious that y is also a function of x, even though no one has actually told you so. When a problem defines x as a function of y, we say it's defined "explicitly." The fact that y is also a function of x isn't really stated, it's implied. The function y is defined "implicitly."

If you took the same situation and said $y = \sqrt{x}$ then y would be explicit, defined as a function of x. The question is really, "which variable is the dependent variable and which is the independent variable?" The one we know is the independent variable, and the other one, the thing we're looking for, is the dependent variable. It's dependent on what goes on in the problem. The independent variable is independent of the math, since it's a known quantity.

Sometimes we know that y depends on the independent variable x, but we can't easily solve the equation so that y is all by itself on one side of the equals sign, with only x's on the other side. In this situation, we throw up our hands and declare "Let y be defined as function of x implicitly by the following equation."

You may be able to find the derivative of y with respect to x implicitly, even though you can't solve the problem for y first. You take advantage of the fact that if two functions are equal, their derivatives are also equal. If (catbox) = (stinko) then (the derivative of catbox *with respect to x*) = (the derivative of stinko *with respect to x*). If you differentiate both sides of the equation with respect to x, the derivatives of both sides are still equal.

Your equation might look like $x^2 + y^2 = 1$. You take the derivative of each side. Whenever you encounter a "y" you remember that y is a function of x, use the chain rule and just use the symbol dy/dx. You get

$$2x + 2y(dy/dx) = 0$$

Then you use algebra to solve for the derivative:

$$2y(\frac{dy}{dx}) = -2x$$

$$\frac{dy}{dx} = \frac{2x}{2y} = \frac{x}{y}$$

That's it. The biggest difference is that the formula for the derivative will probably contain both x's and y's.

The Many Masks of the Derivative

Leibniz and Newton both invented calculus at about the same time. Although their ideas were similar, they used different notations. Both styles of notation have survived and folks have been gleefully adding to them for the last four hundred years.

The result is that there are several ways to symbolically express the same mathematical ideas. Mathematicians use them interchangeably, but specific styles have become linked to specific processes. Your instructor and calculus book may both forget to mention this to you, perhaps because of a post-hypnotic suggestion made by some diabolical cat. You will see a bunch of things that look very different and suspect you've got many new concepts to learn. The truth is, many of the symbols are duplicates and don't represent anything new at all.

148

For example, in algebra we grouped things together in simple parentheses. As you study more complex combinations, it becomes useful to arrange these groups themselves into little families. Because it can be confusing to see several identical parenthesis marks together, they assemble these groups with square brackets. If these are to be joined as families, they are combined within larger flowing brackets.

Braces{Then Square Brackets[Then Parentheses(It starts over in here)]}

The most interesting and potentially confusing is the derivative itself. Leibniz described it as dy/dx and that is still used. When you see that, it means the change in y compared to the change in x, which is exactly what the derivative is. Newton used a dot above the y to mean dy/dt, the derivative with respect to time. Physics folks still use this notation.

Today, we use a little apostrophe after the function sign. If the function is $f(x)$, then the derivative of that function is $f'(x)$ This is the same thing as Leibniz's dy/dx.

Of course, why bother to write out the whole $f'(x)$ when you can simply write f'. Same meaning. It's the derivative of the function.

Likewise, since we're usually interested in y as a function of x, we might describe the function itself as

$$y=f(x)$$

Do not be surprised if you see the derivative of that described as y'. It's the same thing as dy/dx, or f'.

Sometimes they'll abbreviate the derivative with a capital D. Most properly this should be followed by a tiny letter telling us what the derivative is taken with respect to. So D_x means "the derivative with respect to x." This means almost the same thing, with this subtle difference: the capital D tells you you're interested in the derivative of everything that follows it. "$x + D(x+y)^2$" means "x plus the derivative of $(x+y)^2$."

Got that? If $y=f(x)$ you might see the derivative conveyed to you as dy/dx, or as $y'=D_xy$, or y' or f'. A mathematician would probably say it like this:

$$y' = D_x y = D_x f(x) = f' = \frac{dy}{dx}$$

As you embark on your calculus adventure, you'll probably see dy/dx a lot. Gradually these other abbreviations will pop up. The text book authors aren't really trying to confuse you. It's just that different styles have gradually become associated with different aspects of the subject and no one even notices any more that there's a certain redundancy here that might confuse an ordinary person. It certainly makes the calculus books look more intimidating.

When cats see this confusion, they purr happily.

Differentials

As you may have noticed, we're often interested in rates of change. Rates are described mathematically as ratios; that is, things that look like fractions. If we're measuring miles and hours, we end up with some number of miles over some number of hours. When we "simplify the fraction" by completing the division it suggests (divide miles by hours) we get our speed. This is very similar to calculus, but it's not exactly the same. The difference is this: calculus is interested in how a tiny change in one variable, like miles, affects the other variable. In the calculus arena we'd say "a tiny change in miles" over (that is,

150

divided by) "a tiny change in hours." This relationship is the derivative we've learned to love. Until we know the specifics of our problem we describe the tiny changes in miles as dy and the tiny change in time as dx. We call dx and dy "differentials." They represent the infinitesimal "difference" between one spot and another, or one time and another.

It may seem that these differentials dx and dy are married to each other, since we see them holding hands in derivatives so often. But this doesn't mean they don't have lives of their own. Some problems can be solved by focusing on the differentials independently. As we barrel down the highway, if we notice the "tiny change in time" between when one sign flashes past us and when the next one does, we may be able to determine the distance between the two signs.

We can work with differentials because we studied our algebra and remember some basic tricks to manipulate expressions to our taste.

We start by jotting down two different ways to describe a derivative. Since they both describe the same thing, we're pretty safe calling them equal:

$$\frac{dy}{dx} = f'(x)$$

We can multiply both sides of an equation by the same thing and the two sides will still be equal. So why not multiply both sides by dx? This eliminates the "fraction" from the equation altogether, which makes the rest of our algebra tricks easier. Multiply both sides by dx:

$$\frac{dy}{dx}(dx) = f'(x)(dx)$$

151

Which get us this:

$$dy = f'(x)\, dx$$

Remember this formula. You might see it again.

We have certainly changed the equation in flavor, meaning, and appearance. What we have not changed is the equality. We've isolated one of those differentials on the left hand side of the equation because we're interested in figuring out what it represents, not simply what its relationship is to the other one.

Translated into words, this equations says that you can discover how much y changes by multiplying the derivative times the amount of change in x. If the speed is fifty miles per hour (that's the derivative) and one-half hour elapsed between signs, you multiply fifty times one-half to learn that your location changed by 25 miles between signs.

Equations that use differentials are often used to get approximations of continuous functions that would be difficult or impossible any other way.

For example, suppose you wake up in the middle of the night hungry to learn the square root of 9.012. This probably means you spent the evening wrestling with litter-box issues. You know that each bag of the kitty litter your cat prefers will cover an area of precisely 9.012 square feet when spread to the precise depth your cat requires. And, of course, your cat insists on a spacious and perfectly square litter box. No problem if he wanted an area of exactly nine square feet: each side would be exactly three feet long. But how big is each side going to be for the litter box you must create? Without differentials, you'd be forced to make very crude approximations and live with an unhappy cat.

First we translate what we know into a function. There are ✓
several ways we could do this, and part of the fun is figuring out which
one will be useful to us. We decide to call one side of the box *y*, and
call the area of the box *x*. We could say that y^2 equals *x*. Or we could
say that *y* equals the square root of *x*. We decide to say that *y* is a function of *x*, and write it like this:

$$y = f(x)$$

And we know how to define that function as well:

$$f(x) = \sqrt{x}$$

$$\left(y^2 = x \right)$$

Since $y = f(x)$ and $f(x) = \sqrt{x}$ then algebra tells us that

$$y = \sqrt{x}$$

The question becomes "how much does *y* move as *x* moves
from 9 to 9.012? That's exactly what our equation is set up to answer.
We'll multiply the derivative (of the function that describes our situation) times the change in *x* to discover how much *y* changes.

Our formula says "the change in *y* equals the derivative of the
function of *x* times the change in *x*." If we plug our information into
that sentence we get "the change in *y* equals the derivative of the
square root of *x* when *x* equals nine times the change in nine" We use
the power rule to discover that the derivative of the square root is 1/6.
We multiply that times the change from 9 to 9.012. The amount of
change (the differential) is .012, the math is simple and the answer is
.02.

This .02 represents the change in *y* from where we started. We
started with *y* equal to 3 and *x* equal to nine. If *y* started out as 3 and
has changed by .02, we need to make each side of our litter box exactly
3.02 feet long.

Integral Calculus

In differential calculus we tried to extract rate information from two related measurements, say, distance and time. If we were interested in a particular mouse, we made a chart of its location at different times, translated that information into a function and used our calculus processes to determine its speed at any point. When we compare distance and time, that relationship, the first derivative, is speed. A derivative is always a rate, a relationship, like miles per hour.

In integral calculus we try to unwind the implications of a rate over time. We know the mouse's speed, we know how much time has elapsed, now we want to know how far he scurried during that interval.

In differential calculus you start with a function and use your various rules to find the derivative. When you apply your integrating processes to that derivative and discover the original function, your answer is called the anti derivative. You're back where you started.

This is simple if you've got a nice little linear mouse that runs at a steady speed. If he's running along at two miles per hour and he runs for an hour, he'll probably discover a scornful but hungry cat waiting for him exactly two miles down the road.

If we draw a chart of this using speed as the vertical axis and time as the horizontal we get this:

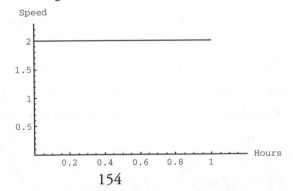

But if the vertical axis represents speed and the horizontal axis is time, where can we find the actual distance traveled during that interval? Remarkably, it is represented by the area. 2 times 1= 2:

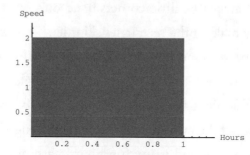

Obviously, you don't need calculus if you're hungry for linear mice. But perhaps your calculus instructor has taken a number of mouse measurements and determined that his furry little prey is moving at a constantly changing speed. By the time you get to class he's reduced this to a chart that looks like this:

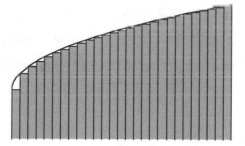

Just like before, the shaded area represents the actual distance traveled. But it's no longer so easy to figure.

One way would be to create lots of rectangles to fill up the area. Since its easy to calculate the area of rectangles, the math is easy. The disadvantage is that you can't make rectangles fill the area perfectly. There will always be little corners unaccounted for. We can get more accurate by making the rectangles skinnier. This reduces the size of the unmeasured sections, but it's still not perfect. We can refine the technique by being more creative with our rectangles, perhaps even making different straight-line shapes to fill the space, but the idea is the same. You can't fill a curved line with straight-edged shapes.

Still, you will spend many fun hours learning this technique, primarily because your calculus instructor had to do it at some point, and, from his perspective, fair is fair. Adding the areas of all these rectangles or other straight edged shapes is called finding the Reimann Sum. Once you have suffered enough, you'll abandon this technique and never really use it again. The exception is for some types of problems that have no elegant solution.

The reason you can abandon it and still catch mice is the Fundamental Theorem of Calculus. This says that finding the shaded area in our diagram is really the reverse process of finding the derivative of the curve along its top, the same way subtraction is the reverse process of addition. The Fundamental Theorem says that integration is the reverse of finding a derivative.

Mathematicians find this beautiful and remarkable, just as you found it remarkable that the shaded area in our example translates into miles.

In integral calculus we start with information that includes a rate and say, "If we were finding the derivative of a function, and this information on the paper was the answer, what function did we start with?"

156

For example, when we find the derivative of x^4 we get $4x^3$.

In integral calculus they give you $4x^3$ (which they now call the integrand) and ask, "what is this the derivative of?" The answer, of course, is x^4. Since this is the exact reverse of finding the derivative it is called finding the anti derivative. The antiderivative is the function you started with. There are several specific procedures to use in accomplishing this, just as there were to find the derivative.

The integrand, or the thing you're integrating, will likely consist of some rate, like 100 miles per hour, times the tiny change in time we're interested in, which we'll describe as dt.

The anti derivative is a function, not necessarily a specific number. You may be reversing a function that can be applied to a whole range of unknowns, but without more information, all you can do is reverse the derivative and get back to the function. That is all you'll need for many purposes. Merely finding the anti derivative is called finding the "indefinite integral." It's indefinite because you don't have all the data. No one has told you during what time period the mouse will be running.

If they have, then you can use the antiderivative to reach a specific number like "23 miles." That answer is called the "definite integral." You know the boundaries of time, say one o'clock and two o'clock, and you want to know exactly how much distance the catmobile covered during that hour.

Just like differential calculus, integration is useful when the rates of change vary over time.

Sometimes you want to add up things within certain boundaries. The symbol for this is the Greek letter sigma, which looks like this

$$\Sigma$$

You put the boundaries of your problem above and below the sign:

$$\sum_{1}^{10} x + 1$$

This means substitute each number between one and 10 for x and add them up. That is, $(1+1)+(2+1)+(3+1)+...(10+1)$

Finding integrals is a similar process, but it isn't exactly the same. We want to add the area of each little rectangle in the shaded portion of our graph. Perhaps we're interested in the time between one and one-thirty.

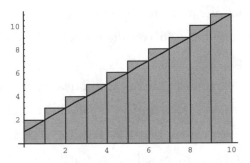

We want to add up all the little skinny rectangles we can create in there to get close to the area. The symbol for this process resembles the letter "s" used for summation and it works in a similar way. It's called the integral sign and it looks like this:

$$\int_{one}^{one-thirty}$$

158

If there are no numbers above or below it, we're looking for the antiderivative. That is, we're looking for the indefinite integral. Perhaps we haven't decided yet when we want lunch, for example, but we still want to keep track of that mouse. Just knowing he's a quadratic mouse, or a cubic mouse, may whet our appetite and we can find that out by knowing the indefinite integral. It will be the function he dances to.

When you see the integral sign followed by something, that something is the derivative of a function times some differential like dt or dx which tells what the independent variable is. Our task is to figure out what that function is. When we see

$$\int_{one}^{one-thirty} \frac{10 \text{ miles}}{1 \text{ hour}} dt$$

We will use our various techniques of integration until we arrive at the answer. Instinctively, you know that the definite integral is five miles. And you thought integral calculus was difficult! You can do it in your head before your first day of class. You instinctively multiplied the integrand, 10 mile per hour, times the period covered, or one-half hour:

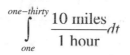

$$\int_{1:00}^{1:30} (10mph)dt = 10mph * \tfrac{1}{2} hour = 5miles$$

In order to confuse you, the cats decided to add a new meaning to the word "limit." In addition to its use in finding increasingly close approximations, they now also use it to refer to the boundaries of the function. In our example, they'd say the "limits" were one and one-thirty. This would be easy enough if they abandoned its previous meaning during integral calculus, but no. That would not be cat-like. When you use the Riemann Sum to find an integral you'll use increasingly skinny rectangles. As the width of each rectangle approaches zero their combined areas will approach the true area. This will be described using the good old limit process just as before. This double meaning will not seem at all weird to your instructor.

The Process

Just as there were specific procedures to find a derivative, there are specific procedures to unwind them. Your first step in differential calculus was figuring out which one to use. Your first step in integral calculus is to figure out which of these procedures was used. Gradually you will develop a knack for this.

Luckily, most of the rules for finding derivatives have matching counterparts in integral calculus. If you decide that the mess you're trying to integrate is the result of someone using the power rule, you've got a specific rule to reverse it. Same for the constant added rule, constant multiple rule and sum rule. It's easy to reverse exponential derivatives if they arrive via the magic of e. Reversing trig functions is as easy as memorizing a little chart.

The product rule requires a different technique called integrating by parts.

There is no exact analog for the quotient rule. However, we can use our vast knowledge and skill at algebra to rearrange those puppies into more tractable beasts.

To reverse the chain rule you use a process called "integration by substitution."

When you begin differential calculus, you may start with the relatively difficult and intimidating difference quotient, just so you gain a good grasp of the math foundation of what you'll do later, and also because dy/dx is a quotient itself. After you've paid your dues and joined the club, you learn the relatively easy rules. Something similar happens in integral calculus. You may be first taught to think of integration as adding the areas of tiny rectangles beneath the curve of a graph (finding the Reimann Sum). The algebra involved in that can be briskly challenging; in fact, forcing someone to endure the process is no longer a legal punishment in most of the civilized world. Only later, when you have suffered enough, will they say, "Oh, by the way, there's a much easier way to do these. Let's look at the Fundamental Theorem."

It's part of the fun initiation process.

The Use of u

The folks who write calculus problems love big, ugly expressions which include various numbers and unknowns besides nicknames for processes like sine and cosine. We may have to treat the whole smelly mess as a single creature when we try to integrate it and it's easy to lose track of what we're doing with all the clutter. It would be impossible to explain what to do in the first place by using any concoction of numbers, variables and nicknames worthy of solving.

So mathematicians have adopted the tradition of describing "a horrible smelly mess of mathematical stuff" with the letter "u". This is not a completely random thing, but it can be confusing.

The letter u is used to represent the "inside function" in an expression. That is, if you have one function (like squaring) operating on another function (like "$x + 4$") we can write it like this:

$$(x + 4)^2$$

The inside function is $x+4$, the outside function is squaring. We replace $x+4$ with the letter u. You probably remember this from the chain rule. On the surface this seems like simple housekeeping, making sure that everything's a little neater. That's certainly one benefit.

But when we replace an expression with "u" we've done something a little more subtle than simple abbreviation. If we were plotting a mouse's activity and calling its location x and comparing that to time, which we called y, our derivative was probably expressed with respect to x. Once we replace all or part of that derivative with u and graph the derivative, we'll be graphing it with respect to u. That sounds trivial, and, in fact, you'll zip past it, but it's really kind of interesting. For example, if we replace $2x$ with u, each unit of u will be twice as large as the original x. Our graph won't be as steep, but the area beneath the curve will be the same when translated into equivalent units. Because the letter u, and the tricks associated with it, refer to an "inside function," it's only used when an expression contains both an inside and outside function.

Using u to replace an important part of the integrand shows up another time which may startle you if you're unprepared. When a book or instructor is explaining one of the strategies of integration in general terms, they may be reluctant to use x to represent the integrand, since the integrand will never be x, but may include an x or two, perhaps even raised to some power. That would be really confusing. So, as they explain the strategies of integration, they will probably use an expression that has u's in it rather than x's. We know that any u they throw at us probably represents a whole cat box full of stuff. Still, your first reaction may be whoa big fella, where did that thing come from? I don't remember any u in the problem. No need to panic. This is part of a natural progression of math communication.

When you learned algebra, a, b, and c represented variables as concepts were explained to you. You quickly realized you weren't ever going to see $(a+b)(a+c)$ on a test. Letters were substitutes for all the things you *would* see on tests, and your instructor hoped you made the connection. Later, when you studied functions, whole processes were abbreviated with a letter f. For many of us, this was a stumbling block. The translation does not seem natural to those of us more used to literature than math. The use of u is similar: it's a generic stand-in for something more complicated or specific. It's unique because it represents not just any old function, but a function that is being acted upon by another function. This may seem very abstract, but it's not really difficult. Think of u as standing for something *ugly*.

Integrating—What Happened to Added Constant?

When you differentiated something that had an added constant, you threw away the constant. Now that you're trying to go backward, that constant has left no trace. Yet, to be accurate, we need to account for it.

We can't know for sure whether there was a constant in that original function or not. So when we integrate, we assume there was. We do whatever we're going to do to get back to the original function, and then we say "plus C." The "C" stands for any constant, even zero.

Backward from the Power Rule

When we found the derivative of a function raised to a power, we moved the exponent in front of the function and reduced the power by one. The derivative of x^4 became $4x^3$.

Working backward from this is pretty obvious: the derivative of $4x^3$ is going to be x^4.

Math folks and cats like to express things like this in a general way that involves letters. The letter they seem to like best for exponents is "n". They describe any number raised to some power as x^n. When they describe the power rule for derivatives they will say:

$$D_x x^n = nx^{n-1}$$

For some people, this is a more concise way to remember what to do.

To reverse the power rule in an integration problem, do the opposite of what you did in the power rule. That is, increase the power by one, then divide by the same number as this new power. Then, of course, you add the unknown constant back in. That is, you'd write "+C" in your answer.

Warning: this process won't work on one integrand, and that single exception drove many mathematicians crazy for years. The exception is the integrand x^{-1}.

Backward from the Constant Multiple Rule

In differentiating a function with some constant multiplied times the rest of the function, we simply moved the constant into the derivative without any fuss. So it only makes sense that, if we're integrating something and we see a constant multiplied times the rest of the integrand, we simply move it back where it started. If you've got 43 times the rest of your integrand, when you're done, your answer will have 43 times the rest of the anti derivative.

Backward from Exponential Functions

If the original function contained an exponential function, and you were looking for the derivative, you probably converted it into an exponential function with a base of e. You did this because the derivative of an exponential function with a base of e is that same function.

It only makes sense that when you're trying to unwind that process, the same logic applies. Because the derivative of e^x is e^x, if you're asked to integrate e^x, you'll quickly answer "e^x." For most integrals involving e^u, you'll be using substitution.

Integrating Trig Functions

All the trig functions are easy to differentiate and some are easy to integrate, which is one reason they are so beloved by calculus-minded cats. There are six trig functions, the derivative of each of them is one of the other trig functions, sometimes with little modifications. Yes, it would certainly take some effort to simply memorize these, just as it takes some effort to memorize your girlfriend's birthday and favorite flower. Those of you who have sustained relationships for more than a year will instantly recognize the lessons learned the hard way and apply them here. In the long run, you will be happier if you just learn them.

The derivative of $\sin(x)$ is $\cos(x)$.
The derivative of $\cos(x)$ is $-\sin(x)$.
The derivative of $\tan(x)$ is $\sec2(x)$.
The derivative of $\sec(x)$ is $\sec(x)\tan(x)$.
The derivative of $\csc(x)$ is $\csc(x)\cot(x)$.
The derivative of $\cot(x)$ is $-\csc2(x)$.

Integration by Substitution

When we're trying to reverse the chain rule, we'll probably replace an expression with u. This is called integration by substitution. It saves us much time and agony.

If you're given a problem like this:

$$\int (x^2 + 3)^3 x\,dx$$

you've got a choice. You could manipulate it with algebra to look like this:

$$\int (x^2 + 3)(x^2 + 3)(x^2 + 3)x\,dx$$

and solve it by multiplying it out and using the reverse of the power rule. But this is a rare situation. Usually you can't manipulate problems to give yourself the option of many calculations before you solve it.

In this case, you could substitute u for (x^2+3) and come up with the answer fairly quickly. The only trick is that you have now moved the problem into a new universe, a universe in which x is not the horizontal axis, it's not what the information is expressed with respect to. We've translated from the world of x to the world of u, per-

166

haps only briefly, but while we're there we'll need to translate *dx* into *du*. Luckily, this is relatively easy. The hard part is remembering to do it.

We are substituting *u* for $x^2 + 3$, and we need to replace the *dx* with an expression involving *du*. We use the differential formula:

$$du = 2xdx$$

But the integral was

$$\int u^3 xdx$$

We solve the differential for *xdx* and get *xdx=du/2*. The new integral is:

$$\int u^3 \frac{du}{2} = \frac{1}{2}\int u^3 du = \frac{1}{2}(\frac{u^4}{4} + C)$$

↑ where does C come from?

or

$$\frac{1}{8}u^4 + C$$

Changing back to the variable *x* our answer is

$$\frac{1}{8}(x^2 + 3)^4 + C$$

This is the antiderivative we wanted.

If we found the derivative of

$$\frac{1}{8}(x^2 + 3)^4 + C$$

we'd get $(x^2 + 3)^3 \, x$

Let's say that our inside function is 12*x* and our outside function is cubing. The integrand looks like this:

$$\int (12x)^3$$

Replacing that inside function with *u* changes the size of the units we're using to measure with. In our example, we've replaced 12 *x*'s with one *u*. This is not rocket science. You could say we just translated from inches to feet. Since there are twelve times as many inches as there are feet in any length, if x = one inch, $12x$ =one foot. If we give $12x$ the name of u, we could also say that u=one foot. In this situation, when we're talking in terms of x, we're talking about inches. When we talk in terms of u, we're using feet. The graph will have a different slope if we're measuring in inches rather than feet, but the area beneath it will be the same. We just need to remember to translate.

Integration by Parts

When we find a derivative by using the product rule we often wind up with something which contains a specific quantity times a derivative plus another specific quantity times a derivative. We can't unwind this mess directly, but we can use the process of "integration by parts" to rearrange the problem to make it easier. And we can repeat the process on the result in hopes of making it even easier. To the delight of cats everywhere, when you repeat the process one too many times you may wind up with something even more difficult to integrate. You have to decide for yourself when to stop. Having choices of strategies may be one reason calculus seems like a glorious fun game to mathematicians and a nightmare for people who crave "one right way" to tackle a problem.

Integration by parts is a way to manipulate each portion of an integrand that resulted from the product rule. It won't actually finish the problem for you. You use it on each "quantity times derivative" section of an integrand. Sometimes that's all you get.

You'll learn to recognize good opportunities to use this technique because the integrand arrives with certain distinctive markings.

It will contain two elements multiplied times each other. One of these will be easy to integrate, the other will be easy to differentiate.

Remember that what we're really looking for, when we integrate, is the original function. If the original function was:

$$(umbrella)(voodoo)$$

then when we use the power rule to find the derivative of this we got:

$$(umbrella')(voodoo) + (umbrella)(voodoo')$$

But that was months (or years) ago during differential calculus. By now you've forgotten both your umbrella and your voodoo. Now, in Integral Calculus, you see something like this:

$$\int(umbrella')(voodoo)dx$$

Notice that it's very easy to find the derivative of voodoo. It's voodoo'. And it's easy to find the integral of umbrella'. It's umbrella. These clues indicate this may be a prime candidate for integration by parts. The reason it's easy to see an integral for one and a derivative for the other is that they are built into the process that created the integrand in the first place. By using the power rule on some function, we got a derivative that fit this profile. What we have now may be enough information to retrace our steps back to the original function.

The process is this: We multiply umbrella times voodoo, then subtract from that answer the integral of voodoo times the derivative of umbrella. In a sentence, that's it. Of course, math books love to abbreviate and use exotic symbols, so they'll express the same thing something like this:

$$\int u\,dv = uv - \int v\,du$$

This handy little formula tells you everything you need to know to integrate by parts.

In this case, u means one of the functions and dv is the differential the other. You'll be substituting real information from various situations, or from problems given to you, into this generalized formula.

Particularly fun is that you may have options about your choice of voodoo. You may be asked, perhaps even required, to integrate something that contains a series of things multiplied together that could be grouped different ways. One rule of thumb is to choose as the element you want to integrate the most complicated expression that you can easily integrate.

In the integral

$$\int x \sin(x)dx$$

a wise choice would be $u=x$ and $dv= \sin(x)dx$. Before you can proceed with the Integration by Parts formula, you must calculate du and v. We find the differential du by noticing the derivative of x is 1. So du $=1\,dx$. We find v by integrating both sides of the differential dv.

$$\int dv = \int \sin(x)dx$$

$$v = -\cos x$$

Now, using the parts formula, we get

$$\int x \sin(x)dx = x(-\cos(x)) - \int -\cos(x)dx = x\cos(x) + \sin(x) + C$$

You *could* have chosen $u=\sin x$ and $dv =xdx$. But the new integral that is the outcome of the integration by parts process would be

$$\int x^2 \cos x dx$$

This one is worse. It has an x^2 in it.

This harmless little kitty of a formula is the meanest beast you will encounter during your first year or so of calculus.

Definite Integrals

The process of unwinding a derivative to locate the antiderivative gets you back to the original function, but it won't give you specific quantities, like "how far did the mouse run?" It will just tell you what function described the mouse's running. If you have specific information ("the mouse ran from two o'clock until two-fifteen") you can use this to get more specific answers. You simply find the antiderivative at two-fifteen and subtract that from the antiderivative at two o'clock. This involves four steps:

1. Find the antiderivative (You can omit the "added c." If you include it, it will get eliminated in step 4 below, when you subtract c.)

2. Plug the upper limit of integration into this antiderivative function

3. Plug the lower limit of integration into this antiderivative function

4. Subtract the lower from the upper.

Little Trick

Say you were asked to find the antiderivative of

(100 miles per hour)(a tiny change in hours)

If that tiny change in hours (which your book will describe as *dx*) is two hours, it's obvious you will travel 200 miles in those two hours.

Plain old algebra tells you that you could double the original function and multiply the result times one half without changing anything.

$$\frac{1}{2}\{(100)(2)(2)\}$$

You can do the same thing to an integrand. If you multiply it times a constant, then multiply the result by the reciprocal of the constant, you haven't really changed anything. If you multiply something by 2, then multiply it times 1/2, you've had some fun, but you're back where you started. In other words, if you see

$$\int x(1+x^2)^2\,dx$$

you can rewrite it as:

how $\frac{1}{2}$ and (2) gets in?

$$\frac{1}{2}\int (2)x(1+x^2)^2\,dx$$

The reason you might be tempted to do such a thing is that it allows you to integrate more directly than some other strategies. You've translated a tricky little problem into one you can solve easily. You'd use $u = 1 + x^2$. So $du = 2x\,dx$. Ultimately, in this case, you'll see that the antiderivative is:

$$\frac{1}{6}(1+x^2)^3 + C$$

how $\frac{1}{6}$?

This really goes over my head!

When You Use Definite Integrals

In calculus class you'll use definite integrals to find the volume or surface area of various three dimensional objects, often created by spinning some shape around its axis. You'll use them to find the center of mass or moment of inertia for some physical object. You'll use

172

them to find the length of a curved line. You might use them to determine how much work a pump does to pump water up to the top edge and over the rim of a container. This is more interesting than it sounds, because the water at the top doesn't have to go as far. If a factory is releasing pollutants at varying rates, you can use the definite integral to "sum up" the total amount of toxic substances released during the period of time your lawsuit covers.

The Waterfall

Imagine you are on a mountainside, and across the valley you see a waterfall. You hunger to learn how far the water falls.

You have a precise stop watch with you, of course. When a log enters the waterfall, you time how long it takes to reach the bottom. Is that enough information?

Yes, if you know how to integrate. The log is being carried along by the water, and falls according to the laws of gravity at an accelerating rate. The rate doesn't change with the weight of the log, and you've memorized the formula for acceleration of falling objects. That rate of acceleration is the derivative of the function that describes its fall; the time the log fell tells you the limits of the integration problem. When you solve it, you'll know exactly how far the log fell, and how big the waterfall is.

This is an example of the practical applications of integral calculus.

Differential Equations

The phrase "differential equation" is counterintuitive. We expect it's something we'll do in differential calculus, but it's really something we do using the techniques we learn in integral calculus. In differential calculus we took known information and worked to find rates; differential equations are just the opposite. They're used when we know the rate of something and want to learn some specific data by working backward from what we do know. Since rates are derivatives, equations we create to describe things that involve rates may contain derivatives as well as functions and other unknowns. These equations are different from simple algebraic ones because they include rate information (derivatives) and also specific functions or quantities. The answer to a differential equation won't be a number, but a function. The unstated question is "what is y?" That is, what function does y stand for? You may get a whole family of functions as your answer. You'll discover the specific quantity you're looking for by using these functions. In integral calculus, you'll reverse the differentiating process to find the antiderivative, which is the original function. Differential equations take that one step further: they incorporate information besides just the function and its derivative.

A cup of coffee will cool at a rate that depends on the difference between the temperature of the coffee and the temperature of the room. It will cool a lot faster in your freezer than it will if you set it on top of a hot car engine. Differential equations are used to figure out how long you'll have to wait so you don't burn your tongue. As the coffee approaches the temperature of the room, the cooling process slows. The difference in two variables affects the rate, but after we describe the rate, now we want to know the specific temperature of the coffee. Our differential equation might include the specific temperature of the room as well as the rate coffee cools at that temperature.

In many situations, the amount of change in some quantity is directly related to the quantity you start with. You'll earn more interest on a big savings account than a small one, for example. The number of kittens born depends on how many mommy and daddy cats you start with. The rate a rumor spreads depends on the number of contacts the folks who know the rumor have with those who don't. Blood flows faster at the center of a blood vessel than it does near the edges; we may know how fast the blood is moving, but it takes some calculus to figure how much blood is actually getting to your brain. We know the rates but not all the quantities. These kinds of problems are solved with differential equations.

People have worked out specific differential equations to model many kinds of situations. The solutions involve finding antiderivatives, usually in some clever way.

It's common to use k to represent a fixed element of a problem, like the interest rate, or the rate of radioactive decay. We might describe growth of a savings account as $y'=ky$, where k represents the interest rate, y represents the amount of interest and y' represents the rate of change of y with respect to time. Some of the known information is derivative (or rate) information, and you want to learn the nonderivative information. You may know speed (derivative information) and you want to know a specific location. Or you may know something about the acceleration (second derivative information) and want to know the specific speed at some point (first derivative information).

The differential equations you'll see in calculus class tend to have solutions that can be expressed with tidy little "closed" formulas. But some free-range differential equations can't be explicitly solved at all. In those cases, you can find a numerical approximation, but it takes some calculating. Before computers became common in the last half of the twentieth century, these had to be done with a pencil and paper and many hours of work.

In the 1940's, scientists tried to figure out what would happen if they exploded an atomic bomb. They were especially interested in learning if their bomb would blow up the whole world. There was no alternative to simply grinding out the math. To solve the problem, the government hired a roomful of women to spend day after day calculating on adding machines. The fact that they used women did not have a sexist connotation: most of the men were busy fighting a war. Even in the 1960's when space travel began, plotting a rocket's course required a ton of calculations. The reason the Apollo missions had to make mid-course corrections was not because the machinery didn't work precisely. It was because there was no elegant way to completely solve the math ahead of time.

Third Semester

Years ago, students first learned integral calculus, then differential calculus. Today, that order is reversed. Most people learn differential calculus in their first semester, and integral calculus in the second semester.

In the third semester of calculus, you typically go back to differential calculus, but now you study functions with more than one independent variable. You'll see things like:

$$z = f(x,y)$$

This may mean you've got a three dimensional mouse that can't be described as handily as a cubic mouse, whose area and volume can be

$z = f(x,y)$ 3 - D EQUATION

both be described as functions of y. Perhaps its nutritional value will be described as

$$z=3x^2 + 4y^2 +2xy$$

First, the derivative rules are adapted into new forms. Many familiar techniques are generalized for multiple variable situations. Of course, there are some new symbols, and more Greek letters casually slipped in. A new concept, the "directional derivative" is thrown in. Now, the steepness of a ski slope depends on which direction you are headed.

Why Study Calculus?

Before steam power ushered in the Industrial Revolution in the 1800's, very few people needed to know calculus. It was useful for aiming cannons and designing cathedrals that wouldn't collapse under their own weight, and plotting the paths of stars and planets. But most people spent more of their time plowing fields, driving ox carts, and cobbling shoes than they did plotting the paths of stars. Calculus was not a hot topic of discussion around many dinner tables.

The power of machines changed that. People had to design steam boilers that wouldn't explode, and pistons that wouldn't fly apart, and looms, cotton gins, locomotives and cars to fine tolerances. It's a lot cheaper to use math to determine optimum sizes and shapes than it is to build a hundred different locomotives and try each one.

A similar thing happened in the 1900's when electricity became important. Electrical components affect each other in many ways, some complementary and some contradictory. When the action of each one is described as a function, calculus predicts how they'll interact without spending a lifetime experimenting in your garage.

In the late twentieth century, as computers became increasingly sophisticated and common, they relieved scientists and engineers of much of the algebraic drudgery of solving the problems calculus could create. This allowed them to design faster and more efficient computers. It also helped them plot pathways to the moon and planets, and analyze the trajectories of distant asteroids which might collide with Earth. Genetic research required the designing of complex equipment; the study of diseases and medicines created huge amounts of data that needed to be analyzed quickly; the military developed increasingly complex weapons needing sophisticated guidance systems and predictable behavior. Business changed, as swirling market forces began to interact with each other on a global basis. Calculus was useful in each of these endeavors.

But, of course, most of us will never design a computer, or aim a "smart bomb," or invest a billion dollars in some convoluted financial instrument. For us, calculus sounds like just another college course that's as distant and irrelevant as 18th century poetry.

The truth is, as the world becomes increasingly complex, calculus becomes increasingly important. One could argue that it's not really necessary to have a computer yet, but the people who understand them have many more options. Calculus is a tool similar to the computer.

If you want to learn business, or biology, or physics, or electronics, or engineering, or medicine, or politics, you will have many more options available to you if you understand calculus. To feel truly educated in the world of the new millennium, you'll need to be as comfortable with it as you are with your computer.

Even if you don't have aspirations along those lines (perhaps you simply love driving that ox cart), learning calculus can sharpen your thought processes, much as learning chess can. Chess has very little practical application for those of us who will never become grand masters, but it develops different thought processes than we exercise in most educational situations. The strategic thinking skills it fosters are invaluable. Many business and military leaders credit learning chess with their ability to think strategically. Similarly, calculus allows us to think in a whole new way. You'll see shadows grow in the afternoon and realize they stretch to the principles of this fluid math, not the rigid linear processes you learned in grade school. You'll think of minima and maxima as you set prices for your products. You'll remember related rates as you design your new home. As you choose your strategies, then use the discipline of algebra to reach conclusions, you are practicing skills that will serve you well for the rest of your life.

The only risk is that you'll have to be a lot more careful around your cat.

THE END

Sine

SOH CAHTOA lived on
KRAKATOA . . .

Sins with cousins: Dan
people get sex in classic
cots.

Please excuse my aunt Sally

oh Heck, another hour of
Algebra